Lecture Notes in Bioinformatics 9957

Subseries of Lecture Notes in Computer Science

More information about this series at http://www.springer.com/series/5381

Eugenio Cinquemani · Alexandre Donzé (Eds.)

Hybrid Systems Biology

5th International Workshop, HSB 2016
Grenoble, France, October 20–21, 2016
Proceedings

 Springer

Editors
Eugenio Cinquemani
Inria Grenoble - Rhône-Alpes
Montbonnot
France

Alexandre Donzé
Department of Electrical Engineering
and Computer Science
University of California at Berkeley
Berkeley, CA
USA

ISSN 0302-9743   ISSN 1611-3349   (electronic)
Lecture Notes in Bioinformatics
ISBN 978-3-319-47150-1   ISBN 978-3-319-47151-8   (eBook)
DOI 10.1007/978-3-319-47151-8

Library of Congress Control Number: 2016952874

LNCS Sublibrary: SL8 – Bioinformatics

Printed on acid-free paper

This Springer imprint is published by Springer Nature
The registered company is Springer International Publishing AG
The registered company address is: Gewerbestrasse 11, 6330 Cham, Switzerland

Preface

This volume contains the papers presented at HSB 2016: the 5th International Workshop on Hybrid Systems Biology held during October 19–20, 2016, in Grenoble (France).

In biology, a "hybrid" is the offspring of two different species. Likewise, in the eco-system of scientific workshops and conferences, the International Workshop on Hybrid Systems Biology (HSB) is the offspring of scientists from two different fields of study, hybrid systems and systems biology. The word "systems" makes the connection between the two, and encompasses the theme common to all works presented at the workshop and published in this volume: The modelling and analysis of complex dynamical behaviors emerging from interacting components of a whole.

Biological systems are complex and, to be useful, models[1] need to capture properties of diverse nature: discrete and continuous, stochastic and deterministic, temporal and spatial, etc. A couple of decades ago, the notion of hybrid system was introduced by computer scientists and mathematicians as a unifying mathematical framework mixing two modelling approaches widely studied at the time: finite-state machines, or automata, and ordinary differential equations. Nowadays, the theory and practice have matured and a hybrid system is widely understood as being a dynamical system mixing continuous and discrete components. The hybrid systems community has grown to become a vibrant interdisciplinary community including logicians, software engineers, applied mathematicians, control theorists, physicist, etc., and, thanks to the many relevant applications, biologists. Different actors bring about different contributions from the respective domains of origin, while at the same time acknowledging the need for complementary expertise and profiting from the pluridisciplinary interaction to enrich their appreciation and solution of scientific and practical challenges.

The Workshop on Hybrid Systems Biology partly emerged as a specialization of hybrid systems to biological case studies, i.e., as a *hybrid systems* biology workshop, but the ambition is for it to become a full-fledged biology workshop, i.e., a hybrid *systems biology* workshop. Here, the adjective "hybrid" means in a broader sense that special attention was paid to true interdisciplinary collaborations and contributions, aimed at reaching beyond some arguably advanced Python scripting performance or abstract methodological proposal of questionable applicability. Hardly any other venue to date can attract and critically evaluate, thanks to the wide and excellent expertise of its Program Committee members, such diverse and multifaceted contributions as on spatial and temporal logics of biological systems, on identifiability and estimation of cellular dynamics, or on the simulation, analysis, and controllability of biochemical pathways, all standing at the frontier of and advancing the fields of computer science, control theory, and of course systems biology.

---

[1] Recall that, quoting George Box, all models are wrong, some are useful.

The first edition of HSB was held in 2012 as a co-located event with CONCUR in Newcastle upon Tyne in the UK. Subsequent editions were HSB 2013, co-located with ECAL 2013 in Taormina (Italy), HSB 2014, co-located with VSL 2014 in Vienna (Austria), and HSB 2015, co-located with CONCUR, QEST, and FORMATS 2015 within the week-long Madrid Meet 2015 in Spain. This year marked a major turning point, because for the first time, the workshop was held as a standalone event. Nevertheless, the workshop attracted as many as 26 submissions. Each submission was reviewed by at least three Program Committee members, with most of them getting four or more reviews. The committee decided to accept 11 high-quality papers, which were organized and presented in four thematic sessions also reflected in this book: Model simulation, model analysis, discrete and network modelling, and stochastic modelling for biological systems. In addition to the oral presentations comprising the main program, the workshop featured an interactive poster and tool demo session, and several high-profile invited talks. The details and full program are available online at http://hsb2016.imag.fr.

The organization of the workshop into such a high-quality scientific and social event would not have been possible without the invaluable help of its general chair, Oded Maler, whose ability to think and reach out to people and topics away from his own scientific center of gravity never ceases to amaze. The editors are also grateful to the help, support, and guidance provided by a highly motivated Steering Committee, in particular, the chairs of the previous editions of HSB, notably David Šafránek, Alessandro Abate, Ezio Bartocci, and Luca Bortolussi. Special thanks go to Sergiy Bogomolov and all Program Committee members who helped us spread the word about HSB 2016, and to all reviewers for providing high-quality evaluations within an extremely tight schedule. Warm thanks to Sophie Azzaro and Catherine Bessière for their invaluable assistance in the practical conference preparation and arrangements. We gratefully acknowledge our sponsoring institutions (Inria, Verimag, UGA, INPG, CNRS, see also the conference website) for their financial and organizational support. Finally, we thank Springer for hosting the HSB proceedings in its *Lecture Notes in Bioinformatics series*, a sub-series of *Lecture Notes in Computer Science*, and Easychair, which lives up to its name by making chairing a workshop such a smooth experience.

August 2016   Eugenio Cinquemani
                                                                    Alexandre Donzé

Organization

General Chair

Oded Maler   CNRS-Verimag, France

Program Committee Chairs

Eugenio Cinquemani   Inria Grenoble – Rhône-Alpes, France
Alexandre Donzé   University of California, Berkeley, USA

Program Committee

Alessandro Abate	University of Oxford, UK
Frank Allgöwer	University of Stuttgart, Germany
Ezio Bartocci	TU Wien, Austria
Grégory Batt	Inria Saclay – Île-de-France, France
Joke Blom	CWI, The Netherlands
Sergiy Bogomolov	IST, Austria
Luca Bortolussi	University of Trieste, Italy
Luca Cardelli	Microsoft Research, Cambridge, UK
Milan Češka	University of Oxford, UK
Eugenio Cinquemani	Inria Grenoble – Rhône-Alpes, France
Pieter Collins	Maastricht University, The Netherlands
Thao Dang	CNRS-Verimag, France
Hidde De Jong	Inria Grenoble – Rhône-Alpes, France
Alexandre Donzé	University of California, Berkeley, USA
François Fages	Inria Saclay – Île-de-France, France
Eric Fanchon	CNRS, TIMC-IMAG, Grenoble, France
Sicun Gao	MIT CSAIL, Cambridge (MA), USA
Radu Grosu	TU Wien, Austria
João Hespanha	University of California, Santa Barbara, USA
Jane Hillston	University of Edinburgh, UK
Sumit Kumar Jha	University of Central Florida, USA
Agung Julius	Rensselaer Polytechnic Institute, USA
Hillel Kugler	Bar-Ilan University, Israel
Oded Maler	CNRS-Verimag, France
Andrzej Mizera	University of Luxembourg, Luxembourg
Chris Myers	University of Utah, USA
Nicola Paoletti	University of Oxford, UK
Ion Petre	Åbo Akademi University, Finland
Tatjana Petrov	IST, Austria

Carla Piazza   University of Udine, Italy
Alberto Policriti   University of Udine, Italy
Guido Sanguinetti   University of Edinburgh, UK
Abhyudai Singh   University of Delaware, USA
P.S. Thiagarajan   Harvard Medical School, USA
Jana Tumova   KTH Royal Institute of Technology, Sweden
Verena Wolf   Saarland University, Germany
Boyan Yordanov   Microsoft Research, Cambridge, UK
Paolo Zuliani   Newcastle University, UK
David Šafránek   Masaryk University, Czech Republic

Additional Reviewers

Ben Sassi,   Laurenti, Luca   Sanwal,
    Mohamed Amin   Lukina, Anna   Muhammad Usman
Demko, Martin   Rocca, Alexandre   Tkachev, Ilya
Forets, Marcelo   Rodionova, Alena   Watanabe, Leandro
Gratie, Diana-Elena   Rogojin, Vladimir   Zhang, Zhen
Halász, Ádám   Samineni, Meher

Contents

Stochastic Modelling

Model Simulation

A Look-Ahead Simulation Algorithm for DBN Models of Biochemical Pathways

Sucheendra K. Palaniappan[1,2(✉)], Matthieu Pichené[1], Grégory Batt[2], Eric Fabre[1], and Blaise Genest[3]

[1] Inria, Campus de Beaulieu, Rennes, France
sucheendra.palaniappan@inria.fr
[2] Inria Saclay - Ile de France, Palaiseau, France
[3] CNRS, IRISA, Rennes, France

Abstract. Dynamic Bayesian Networks (DBNs) have been proposed [16] as an efficient abstraction formalism of biochemical models. They have been shown to approximate well the dynamics of biochemical models, while offering improved efficiency for their analysis [17,18]. In this paper, we compare different representations and simulation schemes on these DBNs, testing their efficiency and accuracy as abstractions of biological pathways. When generating these DBNs, many configurations are never explored by the underlying dynamics of the biological systems. This can be used to obtain sparse representations to store and analyze DBNs in a compact way. On the other hand, when simulating these DBNs, *singular* configurations may be encountered, that is configurations from where no transition probability is defined. This makes simulation more complex. We initially evaluate two simple strategies for dealing with singularities: First, re-sampling simulations visiting singular configurations; second filling up uniformly these singular transition probabilities. We show that both these approaches are error prone. Next, we propose a new algorithm which samples only those configurations that avoid singularities by using a look-ahead strategy. Experiments show that this approach is the most accurate while having a reasonable run time.

1 Introduction

Biological pathway systems are usually modeled by a network of reactions. Their dynamics is governed by mathematical formalisms that are usually either ordinary differential equations (in case of deterministic systems) or some stochastic processes (for non-deterministic systems). Usually, owing to non-linearity and high dimensionality of the system, it is almost always impossible to obtain closed form solutions. Instead, large scale numerical simulations are performed, which are computationally resource intensive. Additionally, there is an inherent limitation of working with biological systems in that usually there is a limited amount of noisy observations of the system. Guided by these considerations, Liu et al., [16] proposed a class of probabilistic graphical models called Dynamic Bayesian

© Springer International Publishing AG 2016
E. Cinquemani and A. Donzé (Eds.): HSB 2016, LNBI 9957, pp. 3–19, 2016.
DOI: 10.1007/978-3-319-47151-8_1

Networks (DBNs) as efficient abstractions of the dynamics defined by the numerical simulations. DBNs have been used to model and study biological systems leading to novel finding in immune system regulation [18].

The basic procedure for abstracting the dynamics of these pathway models as DBNs first consists of discretizing the time and value domain of all the biological species in the model. The state of every biological species of the model corresponds to a discrete-valued random variable in the DBN which represents the discretized concentrations of the associated species. Additionally, it is assumed that each of these random variables are locally dependent on a small subset of other random variables between adjacent time points. This local dependency is dictated by the network structure defined by the underlying mathematical formalism. Having fixed the structure of the DBN, next, assuming a set of initial configurations of the system, a large number of representative numerical simulations are drawn. They are used to then fill the Conditional Probability Tables (*CPTs*) of the DBN by a simple counting strategy. This can be made efficiently using the power of GPUs [15]. More details of this abstraction are described in Sect. 2.2. Once we have constructed the abstraction, instead of analyzing the complex mathematical model, one can study the properties of the biochemical model by analyzing the much simpler DBN model. This significantly reduces the computational burden of repetitive analysis tasks on the original system as shown in [17,18]. Probabilistic verification methods can also be applied [11,15] to further augment the framework.

Analysis on the DBN often transform to the task of computing the probability distributions of the random variables at the different time points. There are two ways of getting these distributions, one by computing the probabilities using a class of approximate inference algorithms [2,12,19,21], and the other by drawing numerous simulations from the underlying DBN and averaging to evaluate the probability distributions. We focus on the latter in this paper. The process of simulation involves starting from a configuration according to the initial distributions of the DBN and drawing a configuration for the next time point according to the conditional probabilities. Contrary to inference, simulations can also be useful when modeling say a tissue of thousands of DBNs, where individual DBNs can have different constraints and cannot be abstracted as a population.

In this paper, we first analyze the specificities of DBNs obtained as abstraction of biological pathways with the method of [16]. We observe that many configurations are never explored by the system, and thus the corresponding entries of the CPTs are not filled. Using case studies, we show that the CPTs are actually very sparse, and we can use this sparsity to encode CPTs in a very efficient way.

However, the fact that many configurations are not explored produces *singularities*, which makes simulation on such DBNs more complex. Typically, drawing a simulation from the DBN involves starting from an initial value assignment over all variables of the DBN (drawn from the initial distribution). Next, for each variable, independently, we pick its value assignment at the next time point according to the distribution dictated by the CPT entries corresponding to its current value assignment. We continue this procedure for all time points of the DBN. In this simulation procedure, singularities appear when a CPT entry for the current value assignment is undefined. This happens when the current value assignment has never been explored, which is possible as values of variables are generated independently (see Sect. 2.3). In the most simple case, one discards these simulations and reruns the simulation procedure until a complete simulation trace is obtained. We show that using such a simulation strategy, which we call *simple simulation*, results in many simulation traces visiting singular configurations. Moreover, it is not very accurate, and quite slow. One way to circumvent the concern of singularities in the DBN is to consider that when the CPT is undefined, all possible future configurations are equally likely. We call this procedure *uniform simulation*. We show that it is fast but also inaccurate.

Our main contribution in this paper is to propose a new simulation strategy, namely *look-ahead simulation*, which generates at every time point tuples of values instead of independently generating these values. The probability transitions are conditioned in such a way to never generate singularities. For efficiency reasons, this is used only when singularities are first encountered, i.e., we follow the simple simulation unless the configuration reached is singular, in which case look-ahead simulation is used. This strategy will be called *adaptive look-ahead* simulation. We show that simulating DBNs using this adaptive strategy is more accurate than the other two. It is also faster than simple simulation when there are lots of singularities. We experimentally evaluate the accuracy and efficiency of these algorithms on three DBN models, two of whom arise from Ordinary Differential Equations (ODE) models of enzyme kinetics and EGF-NGF pathway and one from a hybrid stochastic deterministic model of apoptosis.

Related work: Probabilistic dynamical models arise in many other settings in the study of biochemical networks. In particular, there is considerable work based on Continuous Time Markov Chains (CTMC) [4,5,10,11,13,14,17,22]. Typically, in these studies, every single reaction changing the configuration is executed individually, leading to extremely small time steps. Analysis methods based on Monte Carlo simulations [7–9,11], probabilistic model checking [14] as well as numerically solving the Chemical Master Equation describing a CTMC [6,10] are used to study these continuous time systems. In contrast, we are using DBNs in this paper, which is a discrete time stochastic model, abstracting biological systems using a fixed (larger) time step.

Structure of the paper: The next section discusses DBN models of biological pathways as presented in [16] and also describes the issue of singularities arising in these DBNs with a simple example. Section 3 describes the simple method of simulating the DBN, and the other alternate strategies to circumvent the singularities in the DBN including our improved simulation procedure. Section 4 describes the DBNs produced automatically from pathways that we consider for experiments in this paper. Section 5 reports our results followed by a conclusion.

2   DBN Models of Biological Systems

We begin by introducing notations to present DBNs. The notations are largely adapted from those introduced in [21].

2.1   Dynamic Bayesian Networks (DBNs)

Throughout the sections we fix an ordered set of n random variables $\{X_1, \ldots, X_n\}$ and let i, j range over $\{1, 2, \ldots, n\}$. We denote by $\boldsymbol{X}$ the tuple $(X_1, \ldots, X_n)$. The random variables take values from the discrete set V of cardinality $K < \infty$. We use x_i and v_i to denote a value taken by X_i. We consider DBNs that are time dependent but with a time invariant structure [19]. The underlying structure is unrolled over a finite number of time points, using the same restrictions as in [17].

We will adopt the following notations. $\boldsymbol{x}_I$ will denote a vector of values over the index set $I \subseteq \{1, 2, \ldots, n\}$. It will be viewed as a map $\boldsymbol{x}_I : I \to V$. We will often denote $\boldsymbol{x}_I(i)$ as $\boldsymbol{x}_{I,i}$ or just $\boldsymbol{x}_i$ if I is clear from the context. If $I = \{i\}$ is singleton, and $\mathbf{x}_I(i) = x_i$, we will identify $\mathbf{x}_I$ with x_i. If I is the full index set $\{1, 2, \ldots, n\}$, we will simply write $\boldsymbol{x}$. We call such a tuple $\boldsymbol{x}$ a *configuration*.

A *Dynamic Bayesian Network (DBN)* is a structure $\mathcal{D} = (\mathcal{X}, T, Pa, \{C_i^t\})$ where:

- T is a positive integer with t ranging over the set of time points $\{0, 1, \ldots, T\}$.
- $\mathcal{X} = \{X_i^t \mid 1 \leq i \leq n, 0 \leq t \leq T\}$ is the set of random variables. These variables will be associated with the nodes of the DBN. X_i^t is the instance of X_i at time point t.
- Pa assigns a set of parents to each node and satisfies: (i) $Pa(X_i^0 = \emptyset)$ (ii) If $X_j^{t'} \in Pa(X_i^t)$ then $t' = t - 1$. (iii) If $X_j^{t-1} \in Pa(X_i^t)$ for some t then $X_j^{t'-1} \in Pa(X_i^{t'})$ for every $t' \in \{1, 2, \ldots, T\}$. Thus nodes at the $t - 1$ time point are connected to nodes at the tth time point remains invariant as t ranges over $\{1, 2, \ldots, n\}$.
- C_i^t is the *Conditional Probability Table (CPT)* associated with the node X_i^t specifying the probabilities $P(X_i^t \mid Pa(X_i^t))$.

The regular structure of the DBNs induces the function PA given by: $X_j \in PA(X_i)$ iff $X_j^{t-1} \in Pa(X_i^t)$ for some t. We define $\hat{i} = \{j \mid X_j \in PA(X_i)\}$, so that $P(X_i^t \mid Pa(X_i^t)) = P(X_i^t \mid \mathbf{X}_{\hat{i}}^{t-1})$. Thus $C_i^t(x_i \mid \boldsymbol{u}_{\hat{i}}) = p$ specifies p to be the probability of $X_i = x_i$ at time t given that at time $t - 1$, $X_{j_1} = \boldsymbol{u}_{j_1}, X_{j_2} = \boldsymbol{u}_{j_2}, \ldots, X_{j_m} = \boldsymbol{u}_{j_m}$ with $\hat{i} = \{j_1, j_2, \ldots, j_m\}$.

The semantics of a DBN is the following: from a configuration $\boldsymbol{y}$ at time $t-1$, the probability to move to configuration $\boldsymbol{x}$ at time t is $P^t(\boldsymbol{x} \mid \boldsymbol{y}) = \prod_{i=1}^{n} C_i^t(\boldsymbol{x}_i \mid \boldsymbol{y}_{\hat{i}})$.

2.2   DBN Models as Abstraction of Biological Systems

The dynamics of a pathway are often modeled by a system of equations. For instance, with ODE systems, there is one equation of the form $\frac{dy}{dt} = f(\mathbf{y}, \mathbf{k})$ for

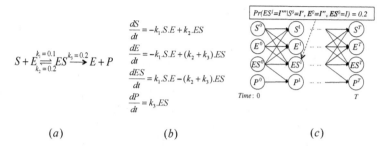

$$\frac{dS}{dt} = -k_1 . S.E + k_2 . ES$$

$$S + E \underset{k_2 = 0.2}{\overset{k_1 = 0.1}{\rightleftharpoons}} ES \overset{k_3 = 0.2}{\longrightarrow} E + P$$

$$\frac{dE}{dt} = -k_1 . S.E + (k_2 + k_3).ES$$

$$\frac{dES}{dt} = k_1 . S.E - (k_2 + k_3).ES$$

$$\frac{dP}{dt} = k_3 . ES$$

(a)   (b)   (c)

Fig. 1. (a) The enzyme catalytic reaction network (b) The ODEs model (c) The DBN approximation for 2 successive time points

each molecular species y, with f describing the kinetics of the reactions that produce and consume y, **y** being the molecules taking part in these reactions and **k** denoting the rate constants associated with these reactions.

Liu et al. developed a dynamic Bayesian network approximation from a system of ODEs [16,17] describing pathway dynamics to compactly store and analyze the dynamics of the original system. Its main features are illustrated by a simple enzyme kinetics system shown in Fig. 1. In the original work, aspects of parameter uncertainty were considered, however, for simplicity, we will not deal with them in this study as it is not the focus of discussion in this paper.

The dynamics of the system is assumed to be of interest only for discrete time points up to a maximal time point. Let us assume that they are denoted as $\{0, 1, \ldots, T\}$. There is random variable Y_i corresponding to every molecular species y_i. The range of each variable Y_i is quantized into a set of intervals $\mathbf{I}^i = \{I_0^i, \ldots, I_{K-1}^i\}$, with K the number of intervals for variable Y_i. The quantized dynamics is intrinsically stochastic, as even for deterministic dynamics (e.g. of an ODE system), it is possible that two distinct deterministic configurations correspond to the same quantized configuration, but their deterministic successors are in distinct quantized configurations. Initial values of the system are assumed to follow a distribution. Initial configurations are sampled according to this distribution, and trajectories are generated by simulating the system from these samples. These trajectories are then compactly approximated as a DBN, treated as an approximation of the dynamics of the system.

For this, the reaction network is used to define the structure (local dependency relation). As discussed before, there is one random variable Y_i^t for each molecular species y_i at time t. The node Y_b^{t-1} will be in $Pa(Y_i^t)$ iff b appears in the mathematical equation of species y_i or if $i = b$.

CPT entries are of the form $C_i^t(v \mid \mathbf{l}) = p$, which says that p is the probability of the value of Y_i falling in the interval I_v^i at time t, given that for each $Y_j^{t-1} \in Pa(Y_i^t)$, the value of Y_j was in $I_{l_j}^j$. We evaluate this probability p through simple counting. For instance, among the generated trajectories, the number of simulations where the value of Y_j falls in the interval $I_{l_j}^j$ at time $t-1$ simultaneously for each $j \in \hat{i}$ is recorded, say J_l. Next, among these J_l trajectories, the number of these where the value of Y_i falls in the interval I_v^i at time t is recorded. If this number is J_v then the empirical probability p is set to be $\frac{J_v}{J_l}$. We refer interested readers to Liu et al.'s work [16] for the details.

We thus have $\sum_{v \in V} C_i^t(v \mid l) = 1$. There is a special case when $J_l = 0$ where we set $p = 0$. This implies that $\sum_{v \in V} C_i^t(v \mid l) = 0$, since all these entries will be null. In this case, we will say that the DBN has *singularities*. This happens when the tuple of intervals corresponding to l has not been visited at time point $t - 1$ by simulating the original system. This is actually the case for a majority of the entries of the DBNs generated from all the biological pathways we tested on (see Sect. 4 and experimental results in Table 1).

2.3   Singularities in DBNs

In this section, we will illustrate how a DBN is generated from a simple system with two variables. We will explain how singularities appear in DBN obtained in such a way, and potential problems arising from these singularities.

Consider a system with 2 variables X_1 and X_2 each of whom take values from the discrete set $\{0, 1\}$. Figure 2 (a) shows the possible configurations of the system for all value assignments of the variables. For instance, the rectangle corresponding to $(0, 0)$ means that the system is at configuration defined by $X_1 = 0$ and $X_2 = 0$ and so on. Figure 2 (a) also shows that the original dynamics of the system forbids it from reaching the configuration $(1, 1)$ i.e., the system cannot reach the configuration with both $X_1 = 1$ and $X_2 = 1$. More precisely, if the system is in configuration $(0, 0)$ at time t, at time $t + 1$, it can stay in the same configuration $(0, 0)$, or go to configuration $(0, 1)$ or $(1, 0)$. However, it cannot go to configuration $(1, 1)$.

Figure 2 (b) shows the DBN construction from the system in Fig. 2 (a), as outlined in the previous subsection. Variables X_1^{t+1}, X_2^{t+1} are dependent on both X_1^t and X_2^t, as shown on the top part of the figure. The bottom part of Fig. 2 (b) shows the conditional probability table for X_1^{t+1} and X_2^{t+1}. For instance, in the

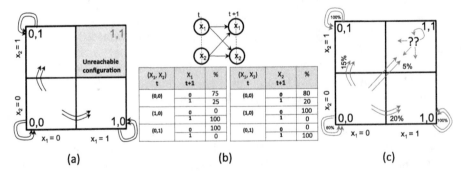

Fig. 2. (a) The dynamics defined by the 2 variable system (b) The assumed DBN model and the CPT entries (c) According to the DBN, every configuration is reachable, including $(1, 1)$ from which no CPT entry is defined

table for X_1^{t+1}, the first column describes the parent configuration tuples, i.e., the values for X_1^t and X_2^t. The second column describes the value X_1^{t+1} can take. The third column represents the corresponding probabilities. The first row describes that if $(X_1^t, X_2^t) = (0,0)$, then X_1^{t+1} taking value 0 (resp. 1) has probability 75 % (resp. 25 %) of happening. This CPT is stochastic: $\sum_{v \in \{0,1\}} C_i^t(v \mid (0,0)) = \sum_{v \in \{0,1\}} C_i^t(v \mid (0,1)) = \sum_{v \in \{0,1\}} C_i^t(v \mid (1,0)) = 1$. Notice that owing to the system dynamics, there is no value in the CPT corresponding to (X_1^t, X_2^t) taking value $(1,1)$, which is a source of singularity in the CPT. Two problems arise from singularities, as shown in Fig. 2 (c). First, from configuration $(X_1^t, X_2^t) = (0,0)$, there is a (small but non null) probability to sample 1 for variable X_1^{t+1} and 1 for variable X_2^{t+1}, and thus to end up in configuration $(1,1)$. This is a problem because configuration $(1,1)$ does not correspond to a behavior of the original system. An additional problem is that from this singular configuration $(1,1)$, there is no defined outgoing probability distribution.

3 Simulation Algorithms

In this section we will first discuss the usual simple algorithm for simulating DBNs. These simulations can potentially encounter singular configurations. In the simple algorithm, such simulations, called *unsuccessful*, are discarded since they may lead to inefficiencies and inaccuracies. We will propose two algorithms to circumvent this issue. The first algorithm simulates CPT entries which are singular by choosing successor configurations uniformly at random. It can be very efficient, although it can be more error prone than simple simulations. Our main contribution is the design of the *look-ahead simulation algorithm*, which samples configurations of the DBNs in such a way to avoid singularities. It thus produces lower unsuccessful simulations, but could be slower.

3.1 Simple Simulation Algorithm

The simple simulation, as shown in Algorithm 1, samples every variable independently. It draws a value w_i for each variable (the **for** loop at line 2) according to the distribution described by $C_i^t(. \mid v_{\hat{i}})$ for v the value assignment which has been sampled at the previous time point (recall that $v_{\hat{i}}$ restricts v to the subset of parents of i). For $t = 0$, w_i is initialized according to the distribution $P(X_i^0)$. The **for** loop at line 10 checks if the new configuration is singular, in which case the simulation is discarded (**Abort** at line 15). Such a simulation is termed *unsuccessful*. We consider for each variable i the set Pre_i^t (line 11) of tuples l of values at time t over the set of parents $\hat{i}$ of node i, and which have been seen in the simulations (that is $\sum_{k \in K} C_i^t(k \mid l) = 1$).

For instance, with the DBN in Fig. 2 (b), from configuration $(0,0)$, there is a small chance to sample $w_1 = 1$ (e.g. $x = rand([0,1]) = 0.9$ which will enter the **while** loop once and set y at 1) and $w_2 = 1$ (for the same reason), in which case the test $w_{\hat{i}} = (1,1) \notin Pre_i^t = \{(0,0), (1,0), (0,1)\}$ (line 11) raises the exception Abort.

Algorithm 1. Simple Simulation of DBNs

1: **for** $t = 0..T$ **do**
2:     **for** $i = 1, ..., n$ **do**
3:         $x \leftarrow rand([0, 1])$
4:         $y \leftarrow 0$
5:         $z \leftarrow C_i^t(y \mid v_{\hat{i}})$ ▷ % For $t = 0$, $C_i^0(y \mid v_{\hat{i}}) = C_i^0(y) = P(X_i^0)$
6:         **while** $x > z$ **do**
7:             $y \leftarrow y + 1, z \leftarrow z + C_i^t(y \mid v_{\hat{i}})$
8:         $w_i \leftarrow y$
9:     **if** $t > 0$ **then**
10:         **for** $i = 1, ..., n$ **do**
11:             $Pre_i^t \leftarrow \{l \in V^{\hat{i}} \mid \sum_{k \in K} C_i^t(k \mid l) = 1\}$ ▷ Set of possible value
    assignments for $\hat{i}$
12:             **if** $w_{\hat{i}} \in Pre_i^t$ **then**
13:                 $v_i \leftarrow w_i$
14:             **else**
15:                 **Abort**

3.2   Uniform Simulation Algorithm

A naive solution to avoid unsuccessful simulations is to assume a predetermined CPT entry $C_i^t(y \mid v_{\hat{i}})$ for singular cases where $\sum_{y \in V} C_i^t(y \mid v_{\hat{i}}) = 0$. For simplicity and compactness of the representation, we assume uniform probabilities: $C_i^t(y \mid v_{\hat{i}}) = 1/K$ (with K the number of values in V) for all $y \in V$ for singular $v_{\hat{i}}$ at time t. In this case, for every t, i and every $v_{\hat{i}}$, we have $\sum_{y \in V} C_i^t(y \mid v_{\hat{i}}) = 1$.

The *uniform stimulation algorithm* thus uses the simple simulation of Fig. 1, however with fully defined CPTs. As a result of this, we no longer have the issue of drawing unsuccessful samples. The **for** loop from line 10 to 15 is unnecessary and hence removed. Although this results in speed-up of drawing effective simulations (all samples are effective), the underlying dynamics can be less accurate, as the information that l has not been sampled is lost with this uniform simulation.

For instance, with the DBN in Fig. 2 (b), from configuration $(0, 0)$, there is a small chance to draw configuration $(1, 1)$ (same as with the simple simulation). Now, using the uniform algorithm, simulations from $(1, 1)$ are not discarded, and instead the system can go to any configuration of $\{(0, 0), (1, 0), (0, 1), (1, 1)\}$, with a 25 % of chance for each.

3.3   Look-ahead Simulation Algorithm

In this section we propose a more involved simulation, called *look-ahead simulation*. The main idea is to sample tuples of values v representing the configuration at time t, according to the configuration y at time $t - 1$ and the corresponding CPTs, but conditioned to the fact that v has been seen at time t at least once during the original simulations of the system. Computing exactly this conditional probability is however not possible in general as that would require to recompute on the fly the probability for each of the many possible tuples v.

Algorithm 2. Look-ahead simulation of DBNs

1: $W[1, ..., n] = \{-1, ..., -1\}$ ▷ $W[i] = -1$ means that the value of variable i is not yet set
2: **for** $i = 1, ..., n$ **do**
3:     $Pre_i^t(W) = \{l \in V^{\hat{i}} \mid \sum_{k \in K} C_i^t(k \mid l) = 1 \wedge \forall m$ with $W[m] \neq -1, l[m] = W[m]\}$
4:     $s \leftarrow 0$
5:     **for all** $l \in Pre_i^t(W)$ **do** ▷ Compute the probabilities for value assignments in $Pre_i^t(W)$
6:         $p \leftarrow 1$
7:         **for all** $j \in \hat{i}$ **do**   ▷ Compute probabilities of l
8:             $p \leftarrow p \cdot C_i^t(l[j] \mid v_{\hat{j}})$
9:         $Prob[l] \leftarrow p$
10:         $s \leftarrow s + p$ ▷ Sum of probabilities in $Pre_i^t(W)$
11:     **if** $s = 0$ **then** Abort   ▷ No configurations are consistent with W
12:     $x \leftarrow rand([0, s])$
13:     $z \leftarrow 0$
14:     **for all** $l \in Pre_i^t(W)$ **do**   ▷ Pick a l according to $Prob$
15:         $z \leftarrow z + Prob[l]$
16:         **if** $x \geq z$ **then**   ▷ The current l has been chosen for variables in $\hat{i}$
17:             **for all** $j \in \hat{i}$ **do**
18:                 $W[j] \leftarrow l[j]$
19:             **break**
20: **for** $i = 1, ..., n$ **do**
21:     $v_i \leftarrow W[i]$

Instead, for efficiency reasons, we over-approximate this set of configurations. We fill up iteratively a partial value assignment W remembering the values of variables which have been already set ($W[i] = -1$ if i has not yet been set). The i-th iteration, focused on the variable i, assigns values for parents of i, that is for variables in $\hat{i}$. For this, we consider the set $Pre_i^t(W)$ which only retains value assignments $v_{\hat{i}}$ for $\hat{i}$ at time t which have been seen in the simulations and which are consistent with W.

Notice that by (inductive) construction of $v_{\hat{j}}$, $C_i^t(l[j] \mid v_{\hat{j}})$ is well defined at line 8. In order to justify Algorithm 2, we prove that it is equivalent to Algorithm 1 in an ideal case where Algorithm 1 is correct. In particular, in that case, the distribution does not depend upon the order in which variables are considered in Algorithm 2.

Theorem 1. *In the case where there is no singular configurations, Algorithm 2 generates the same distribution over configurations W than Algorithm 1.*

Proof (Sketch of). The probability of choosing W for Algorithm 1 is the product $\prod_j C_j^t(W[j] \mid v_{\hat{j}})$. We show by induction on the number of variables that W is also generated with this probability by Algorithm 2. If there are no variables, this is trivial.

Since there are no singular configurations, $Pre_i^t(W) = \{l \in V^{\hat{i}} \mid \forall m$ with $W[m] \neq -1, l[m] = W[m]\}$. We will show the property is true for W restricted to variables $I' = \hat{i} \cup \{j \mid W[j] \neq -1\}$, which we denote by $W|_{I'}$. Let $J = \{j \in \hat{i} \mid W[j] \neq -1\}$. It is easy to see that at line 10 of Algorithm 2, s will be equal to $s = \prod_{j \in J} C_j^t(W[j] \mid v_{\hat{j}})$, because for all $j \in \hat{i} \setminus J$, $\sum_{l[j]} C_i^t(l[j] \mid v_{\hat{j}}) = 1$ (no singular configurations). Hence choosing $W|_{I'}$ will be done with probability $\frac{\prod_{j \in \hat{i}} C_j^t(W[j] \mid v_{\hat{j}})}{s}$ (to select $W_{\hat{i} \setminus J}$ at line 14), times $Prob(W|_{\{j \mid W[j] \neq -1\}})$ (to have selected $W|_{\{j \mid W[j] \neq -1\}}$ after $i - 1$). By induction hypothesis, the second term is $\prod_{j \mid W[j] \neq -1} C_{\hat{j}}^t(W[j] \mid v_{\hat{j}})$. The first term can be simplified into $\prod_{j \in \hat{i} \setminus J} C_j^t(W[j] \mid v_{\hat{j}}) \frac{s}{s}$. This proves the result by induction. □

The set of value assignments $v_{\hat{i}}$ for $\hat{i}$ which have been explored is small (< 1000, see Table 1 in Sect. 5.1), compared to the set of visited value assignments v for all variables. In order to set a value assignment $v_{\hat{i}}$ for variables in $\hat{i}$, we draw one value assignment in $Pre_i^t(W)$, according to its probability w.r.t. the sum of probabilities of all value assignments in $Pre_i^t(W)$ (that is the probability of drawing a value assignment in $Pre_i^t(W)$ is the same as for the simple simulation, but conditioned to the fact that a value assignment of $Pre_i^t(W)$ is drawn). If all variables of W can be set in this way, then by definition we have that W is not a singular configuration.

This look-ahead algorithm can however be quite slow because of the overhead of recomputing probabilities conditioned to sampling only non singular configurations. To speed up the process, we use an *adaptive* procedure, which first uses the simple simulation, and only at the time point when a singularity is encountered uses the look-ahead algorithm to compute a new value assignment at this time point.

Notice that it may still be the case that the look-ahead algorithm Abort (line 11), in case where there is no value assignments in $Pre_i^t(W)$ (that is $s = 0$).

For instance, in the example from Sect. 2.3, $Pre_1^{t=1}(W) = Pre_1^{t=1}(\{-1, -1\}) = \{(0,0),(1,0),(0,1)\}$. Starting from configuration $(0,0)$ at time $t = 0$, we compute $Prob[(0,0)] = 60\%$, $Prob[(0,1)] = 15\%$, and $Prob[(1,0)] = 20\%$. We have $s = 95\%$, and we sample $(0,0)$ with probability $60/95$. Notice that $(1,1)$ will never be generated. Assume that $(0,0)$ is drawn (e.g. $x = rand([0,1]) = 0.5$). Now, $W = \{0,0\}$, and we have $Pre_2^{t=1}(\{0,0\}) = \{(0,0)\}$. It gives $Prob[(0,0)] = 60\%$, $s = 60\%$ and thus $(0,0)$ is sampled with probability $60/60 = 1$. On this simple example, there is no unsuccessful trajectory using the look-ahead simulation.

4   DBN Models Considered in the Experiments

In this section, we will introduce the case-studies that we use to construct DBNs and test the different DBN simulation algorithms we discussed in the previous section.

4.1   Enzyme Catalytic Kinetics

The enzyme catalytic system is shown in Fig. 1. It describes a typical mass action based kinetics of the binding (ES) of enzyme (E) with substrate (S) and its subsequent catalysis to convert the substrate into product (P). The value space of each variable is divided into 5 equal intervals. The time scale of the system is 10 minutes which was divided into 100 time points. To fill up the conditional probability tables, $5 \cdot 10^5$ trajectories were generated from the ODE model.

4.2   EGF-NGF Pathway

The EGF-NGF pathway describes the behavior of PC12 cell to EGF or NGF stimulation [3]. The ODE model of this pathway is available in the BioModels database [20]. It consists of 32 differential equations (one for each molecular species). As before, the value domains of the 32 variables were divided into 5 equal intervals. The time horizon of each model was set at 10 minutes which was divided into 100 time points. To fill up the conditional probability tables, $5 \cdot 10^5$ trajectories were generated by simulating the ODE model.

4.3   Apoptosis Pathway

TNF-related apoptosis-inducing ligand (TRAIL) is a protein that is known to induce apoptosis in cancer cells and hence has been considered as a target for anti-cancer therapeutic strategies. However, biological observations suggest that in a population of cells, application of TRAIL leads to only fractional killing of cells (for instance, 70 % of deaths after 8 h for 250 ng/ml of TRAIL) and also a time dependent evolution of cell resistance to TRAIL. Several models have

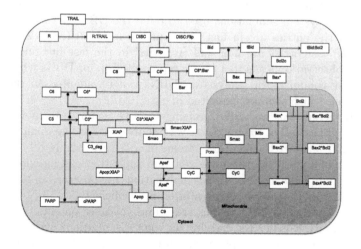

Fig. 3. Apoptosis pathway

been proposed to understand this mechanism, specifically an Hybrid Stochastic Deterministic (HSD) model [1] which combines a deterministic signal transduction model (modeled as ordinary differential equations (ODEs)), and a stochastic model for protein synthesis fluctuations (factoring cell-to cell variability). The model consists of 58 protein variables, the time horizon of the model was studied over an 8 hour period which was divided into 45 time points. To fill up the conditional probability tables, $5 \cdot 10^5$ trajectories were generated by simulating the HSD model (Fig. 3).

5   Experiments

We now evaluate the different DBN simulation algorithms on the DBNs generated for the systems described before. We measure the % of effective samples, mean error, and time taken per effective sample to assess the different approaches. We also discuss the aspect of sparsity in CPTs for all of these DBNs.

5.1   Sparse CPTs

For each of the DBNs that we constructed, Table 1 shows the base statistics in terms of the number of variables and time-points considered for them. For a DBN, the maximal number of CPT entries corresponding to any variable of the DBN is given by K^{N+1} where N the maximal number of parents any node of the DBN can have and K is the cardinality of that node. For instance, in the enzyme kinetics system, $N = 3$ (see Fig. 1). Similarly, $N = 5$ for EGF-NGF pathway and $N = 8$ for apoptosis pathway. In all our case studies $K = 5$ for all variables, the number of discrete intervals we assume. If represented explicitly, these entries potentially make for a large CPT table, whose size is reported in MB. These numbers are shown in Table 1.

  As explained in Sect. 2.3, some of these entries may be 0 since they have not been visited by system simulations used to build the DBN. Actually, we notice that most of the entries are in fact null: only at most 10 (resp. 28 and 851) out of the K^{N+1} entries are non null (reported in the last but one column). This simple observation leads to a sparse representation for DBNs which is very compact (last column). All analysis on DBNs can be adapted to this sparse representation, which was not exploited in e.g. [16,17].

  Over all the systems we tested, ranging from very simple to more complex ones, we found that all of them exhibit very sparse CPTs. The main reason for this can be attributed to the fact that parents of nodes are usually very correlated and capture the underlying dynamics well, and while they can individually take all of the 5 values, only very few of the combinations are observed in the systems, which explain why the sparse representation is so efficient.

  Notice that although CPT entries are sparse (because parents of nodes are highly correlated), it is not the case for the joint distributions over all the variables. For instance, in the apoptosis pathway, out of 200000 simulations, 95 % of the simulations resulted in distinct joints, that is very few end up having the

Table 1. Sparsity in CPTs

Case study	Number variables	Number time points	Max number CPT entries per value assignment	Explicit DBN size (MB)	Max CPT entries $\neq 0$ per value assignment	Sparse DBN size (MB)
Enzyme	4	100	125	1.1	10	0.049
EGF-NGF	32	100	3125	434.4	28	0.605
Apoptosis	58	45	390625	> 20000	851	1.5

same joints (at most 5 % reach the same joints). This implies that explicitly keeping every non null joints over all variables is not feasible, although it is a practical once restricted to parents of a variable (which is what the DBN encodes well with a sparse encoding).

5.2   Evaluation of Simulation Algorithms

We designed tests to compare the different simulation algorithms presented in Sect. 3 (called simple, uniform and adaptive (for the adaptive look-ahead algorithm) in the Tables 2 and 3). We evaluated them on the different DBNs generated from the biological systems described in Sect. 4. For all case studies reported here we perform experiments by drawing 10000 samples from the underlying DBN and analyzing them.

We mainly measure and compare (a) the % of effective runs for the algorithms (not reported for uniform simulation as it is always 100 %), (b) the time needed for each effective simulation, and (c) the difference between outputs of these three algorithms and the original dynamics (Enzyme, EGF-NGF, Apoptosis), in terms of mean error of the different trajectories. Mean error is computed as mean over the weighted difference between the marginals computed by the DBN algorithm and marginals of the quantized traces of the original system of each variable.

All results are summarized in Table 2. For the enzyme kinetics system, almost all simple simulations are effective (98 %), and so all the three algorithms show no tangible difference, with a very good agreement with the dynamics of the original biological pathway.

Table 2. Comparing uniform, simple and adaptive look-ahead simulations

Case study	Effectiveness		Time per effect. sample			Mean error uniform		
	simple	adaptive	uniform	simple	adaptive	uniform	simple	adaptive
Enzyme	98 %	100 %	1.81 ms	2.33 ms	2.26 ms	0.454 %	0.347 %	0.429 %
EGF-NGF	36 %	100 %	20.3 ms	30.4 ms	26.8 ms	2.22 %	1.63 %	1.30 %
Apoptosis	3.1 %	83 %	8.46 ms	121 ms	68.0 ms	14.7 %	6.82 %	4.47 %

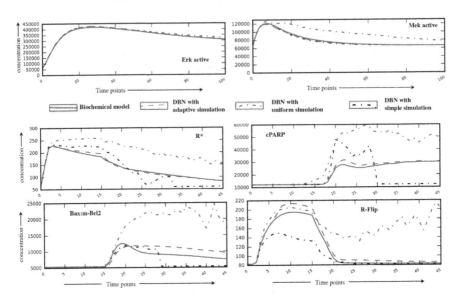

Fig. 4. Mean value of trajectories for selected species in the EGF-NGF pathway (top 2) and in the apoptosis pathway (bottom 4), as simulated by the biochemical model (solid green), the DBN using the simple simulations (black), adaptive look-ahead simulations (red), and uniform simulations (blue) (Color figure online)

For the EGF-NGF case, there was a clear percentage of unsuccessful simple simulations (64 %), and the three algorithms give considerably different results. The adaptive look-ahead algorithm is overall more faithful than the simple simulation while being as fast, however, the uniform algorithm is faster but less accurate (70 % more errors than the adaptive look-ahead algorithm). The mean error may seem small because for most proteins, there is a very low error (e.g. ERK active in Fig. 4 top left), although tangible difference could be seen (especially using the uniform simulation) on few proteins (e.g. Mek active in Fig. 4 top right).

For Apoptosis, the number of effective simple simulations was very low (3.1 %), and the errors of our algorithms w.r.t. the biological system were larger, as shown in Fig. 4 (bottom). The simple algorithm was the slowest and quite inaccurate, due to very few effective samples. The uniform algorithm was very fast, while being very inaccurate, with very large errors on some proteins, as shown in Fig. 4 (bottom).

Overall, the adaptive look-ahead algorithm succeeds in bringing up to 24 times (Apoptosis in Table 2) more effective samples compared with the simple simulation. It fails to reach 100 % only in the extreme case where most simple simulations fail. The adaptive look-ahead simulation clearly improves the accuracy of the simulation. Last, runtime is not impacted by choosing adaptive look-ahead over simple simulation. Instead, in extreme case, adaptive look-ahead simulation can outperform simple simulations in the number of effective samples being simulated per unit of time (Table 2).

Table 3. Percentage dead cells for different simulation strategies

Apoptosis	HSD model	Uniform simulation	Simple simulation	Adaptive simulation
dead cells	70 %	99.4 %	99.7 %	73 %

There might be specific cases where the performance of simulation algorithms can become very important, for instance, when the simulations can have different lengths (e.g. a simulation is truncated when a cell dies in the apoptosis pathway). In this case, a DBN simulation can lead to configurations of the system representing cell death. If the simulation reached some of these configurations, then the cell is declared dead and no further simulation is performed. This also means that an alive cell will have simulation traces equal to the number of time points considered. Consequentially, owing to the limitations of the simple simulation, generating alive cells is considerably harder. This is evident in Table 3, where the simple and uniform strategy produce > 99 % cell death, while the original system dynamics [1] predicts a 70 % death rate.

The simple strategy also has an extremely low efficiency (3.1 %): almost all effective simple simulations of the apoptosis pathway end up dead (99.7 %). Uniform simulations give very inaccurate results as well (99.4 % vs 70 % of dead cells). On the other hand, using the look-ahead simulation, most of the simulations are effective (83 %). We also obtain a death rate close to the one of HSD model (73 % instead of 70 %).

6 Conclusions

In this paper, we have highlighted and systematically characterized singularities observed in DBNs obtained as abstractions of biological pathway dynamics [17] following the method of [16]. The main characteristic of these DBNs is that many entries of the CPTs are null. We show that this can be used to vastly improve the representation of DBNs using sparse encoding. However, it is also a source of inaccuracy when simulating these DBNs. Compared with the usual algorithm which is used to simulate DBNs, we propose two different simulation algorithms. First, a uniform simulation procedure which is faster but also less accurate. This should be preferred when time as a resource is limited and a crude evaluation is sufficient. More importantly, we developed a new adaptive look-ahead simulation which is comparable in run time or faster than simple simulation and also is the most accurate, as it produces very limited number of unsuccessful simulations compared to simple simulations (up to 24 times less). In terms of future work, we plan to develop a class of models more accurate than DBNs, that would encode probabilities of tuples of variables in CPTs rather than over single variables.

Acknowledgments. This work was partially supported by ANR projects STOCH-MC (ANR-13-BS02-0011-01) and Iceberg (ANR-IABI-3096).

References

1. Bertaux, F., Stoma, S., Drasdo, D., Batt, G.: Modeling dynamics of cell-to-cell variability in TRAIL-induced apoptosis explains fractional killing and predicts reversible resistance. PLoS Comput. Biol. **10**(10), 14 (2014)
2. Boyen, X., Koller, D.: Tractable inference for complex stochastic processes. In: UAI-98, pp. 33–42 (1998)
3. Brown, K.S., Hill, C.C., Calero, G.A., Lee, K.H., Sethna, J.P., Cerione, R.A.: The statistical mechanics of complex signaling networks: nerve growth factor signaling. Phys. Biol. **1**, 184–195 (2004)
4. Calder, M., Vyshemirsky, V., Gilbert, D., Orton, R.J.: Analysis of signalling pathways using continuous time Markov chains. In: Priami, C., Plotkin, G. (eds.) Transactions on Computational Systems Biology VI. LNCS, vol. 4220, pp. 44–67. Springer, Heidelberg (2006). doi:10.1007/11880646_3
5. Danos, V., Feret, J., Fontana, W., Harmer, R., Krivine, J.: Rule-based modelling of cellular signalling. In: Caires, L., Vasconcelos, V.T. (eds.) CONCUR 2007. LNCS, vol. 4703, pp. 17–41. Springer, Heidelberg (2007)
6. Didier, F., Henzinger, T.A., Mateescu, M., Wolf, V.: Approximation of event probabilities in noisy cellular processes. Theor. Comput. Sci. **412**, 2128–2141 (2011)
7. Donaldson, R., Gilbert, D.: A model checking approach to the parameter estimation of biochemical pathways. In: Heiner, M., Uhrmacher, A.M. (eds.) CMSB 2008. LNCS (LNBI), vol. 5307, pp. 269–287. Springer, Heidelberg (2008)
8. Fages, F., Rizk, A.: On the analysis of numerical data time series in temporal logic. In: Calder, M., Gilmore, S. (eds.) CMSB 2007. LNCS (LNBI), vol. 4695, pp. 48–63. Springer, Heidelberg (2007)
9. Grosu, R., Smolka, S.A.: Monte Carlo model checking. In: Halbwachs, N., Zuck, L.D. (eds.) TACAS 2005. LNCS, vol. 3440, pp. 271–286. Springer, Heidelberg (2005)
10. Henzinger, T.A., Mateescu, M., Wolf, V.: Sliding window abstraction for infinite Markov chains. In: Bouajjani, A., Maler, O. (eds.) CAV 2009. LNCS, vol. 5643, pp. 337–352. Springer, Heidelberg (2009)
11. Jha, S.K., Clarke, E.M., Langmead, C.J., Legay, A., Platzer, A., Zuliani, P.: A Bayesian approach to model checking biological systems. In: Degano, P., Gorrieri, R. (eds.) CMSB 2009. LNCS, vol. 5688, pp. 218–234. Springer, Heidelberg (2009)
12. Koller, D., Friedman, N.: Probabilistic Graphical Models - Principles and Techniques. MIT Press, Cambridge (2009)
13. Kwiatkowska, M.Z., Norman, G., Parker, D.: Probabilistic model checking for systems biology. Symbolic Systems Biology, Jones and Bartlett (2010)
14. Kwiatkowska, M., Norman, G., Parker, D.: PRISM: probabilistic symbolic model checker. In: Field, T., Harrison, P.G., Bradley, J., Harder, U. (eds.) TOOLS 2002. LNCS, vol. 2324, pp. 200–204. Springer, Heidelberg (2002). doi:10.1007/3-540-46029-2_13
15. Liu, B., Hagiescu, A., Palaniappan, S.K., Chattopadhyay, B., Cui, Z., Wong, W.-F., Thiagarajan, P.S.: Approximate probabilistic analysis of biopathway dynamics. Bioinformatics **28**(11), 1508–1516 (2012)
16. Liu, B., Hsu, D., Thiagarajan, P.S.: Probabilistic approximations of ODEs based bio-pathway dynamics. Theor. Comput. Sci. **412**, 2188–2206 (2011)
17. Liu, B., Thiagarajan, P.S., Hsu, D.: Probabilistic approximations of signaling pathway dynamics. In: Degano, P., Gorrieri, R. (eds.) CMSB 2009. LNCS, vol. 5688, pp. 251–265. Springer, Heidelberg (2009)

18. Liu, B., Zhang, J., Tan, P.Y., Hsu, D., Blom, A.M., Leong, B., Sethil, S., Ho, B., Ding, J.L., Thiagarajan, P.S.: A computational, experimental study of the regulatory mechanisms of the complement system. PLoS Comput. Biol. **7**(1), e1001059 (2011)
19. Murphy, K.P., Weiss, Y.: The factored frontier algorithm for approximate inference in DBNs. In: UAI 2001, pp. 378–385 (2001)
20. Le Novere, N., Bornstein, B., Broicher, A., Courtot, M., Donizelli, M., Dharuri, H., Sauro, H., Li, L., Schilstra, M., Shapiro, B., Snoep, J., Hucka, M.: Biomodels database: a free, centralized database of curated, published, quantitative kinetic models of biochemical and cellular systems. Nucleic Acids Res. **34**, D689–D691 (2006)
21. Palaniappan, S.K., Akshay, S., Genest, B., Thiagarajan, P.S.: A hybrid factored frontier algorithm. TCBB **9**(5), 1352–1365 (2012)
22. Hlavacek, W.S., Faeder, J.R., Blinov, M.L., Posner, R.G., Hucka, M., Fontana, W.: Rules for modeling signal-transduction systems. Sci. STKE **344**, re6 (2006)

Accelerated Simulation of Hybrid Biological Models with Quasi-Disjoint Deterministic and Stochastic Subnets

Mostafa Herajy[1(✉)] and Monika Heiner[2]

[1] Department of Mathematics and Computer Science, Faculty of Science,
Port Said University, Port Said 42521, Egypt
mherajy@sci.psu.edu.eg
[2] Computer Science Institute, Brandenburg University of Technology,
Postbox 10 13 44, 03013 Cottbus, Germany
http://www-dssz.informatik.tu-cottbus.de

Abstract. Computational biological models are indispensable tools for in silico hypothesis testing. But with the increasing complexity of biological systems, traditional simulators become inefficient to tackle emerging computational challenges. Hybrid simulation, which combines deterministic and stochastic parts, is a promising direction to deal with such challenges. However, currently existing algorithms of hybrid simulation are impractical for implementing real and complex biological systems. One reason for such limitation is that the performance of hybrid simulation not only relies on the number of stochastic events, but also on the type as well as the efficiency of the deterministic solver. In this paper, a new method is proposed for improving the performance of hybrid simulators by reducing the frequent reinitialisation of the deterministic solver. The proposed approach works well with models that contain a substantial number of stochastic events and higher numbers of continuous variables with limited connections between the deterministic and stochastic regimes. We tested these improvements on a number of case studies and it turns out that, for certain examples, the amended algorithm is ten times faster than the exact method.

Keywords: Accelerated hybrid simulation · Deterministic and stochastic simulation · Dependency graph · Computational modelling

1 Introduction

Hybrid simulation of biological reaction networks [11,23,28], which combines stochastic, deterministic, as well as other simulation algorithms, represents one important direction to deal with the recent challenges due to the increasing complexity of biological models [11,16,19]. It can successfully execute complex systems where molecular fluctuations drive the model results, but (discrete) stochastic simulation fails to produce any results in reasonable time. Moreover,

© Springer International Publishing AG 2016
E. Cinquemani and A. Donzé (Eds.): HSB 2016, LNBI 9957, pp. 20–38, 2016.
DOI: 10.1007/978-3-319-47151-8_2

hybrid simulation provides an elegant tool to control the accuracy as well as the simulation efficiency by enabling the modeller to select an appropriate simulator to execute individual parts of a reaction network [14,18,19]. In addition, it can effectively deal with stiff models where reaction species and/or the reaction rate constants exhibit a considerable variation [14]. In this paper, we confine ourselves to a type of hybrid simulation that combines the deterministic and stochastic regime [11]. The deterministic part is executed by solving a system of ordinary differential equations (ODEs) and the stochastic part by considering all reactions as discrete and stochastic processes.

The hybrid simulation algorithm works by first partitioning a reaction network into two subnets: stochastic and continuous [11]. Afterwords, each subnet is executed using the corresponding simulator. Usually, the two regimes are not completely disconnected from each other. Instead, they share interfacing species/reactions that interconnect the execution of the two subnets. Thus, a synchronisation module is highly required to transfer the control between the two simulators. There are two main approaches to synchronise the operation of the internal simulators: approximate [11,23] and exact [1,28].

An approximate synchronisation starts by solving a system of ODEs until a discrete event occurs. The time at which the discrete event occurs is roughly located by repeatedly trying to take an appropriate time step. On the contrary, the exact synchronisation approach finds the correct time where the discrete event is scheduled to occur by incorporating an additional ODE to the deterministic regime representing a time transformation. In either of the two approaches, the stochastic module fires the occurring reaction and gives the control back to the deterministic solver.

Furthermore, approximate synchronisation does not provide sufficient accuracy as it does not capture all stochastic events. It aims to let the ODE solver to take a longer step size and subsequently improves the simulator performance. In contrast, the exact synchronisation approach can successfully locate all stochastic events in the discrete part. However, its performance may be prohibitively slow, which in many examples renders the hybrid simulation ironically slower than the stochastic one. There are many factors that contribute to such behaviour. One of the major reasons is the type of ODE solver.

A hybrid simulator requires an ODE solver that takes flexible steps such that the exact location of the stochastic event is precisely captured. Unfortunately, such type of ODE solver makes a significant effort to initially select the appropriate step size that satisfies certain accuracy criteria. However, hybrid simulation necessitates the reinitialisation of the ODE solver each time the simulator switches from the discrete to the continuous regime. This step is imperative such that the deterministic solver can account for the discontinuities due to the firing of the stochastic reactions. Consequently, as the number of stochastic events increases, the performance of the hybrid simulation rapidly decreases.

In this paper, we propose an approach for improving the performance of hybrid simulation by significantly reducing the number of times the ODE solver is reinitialised. We utilise two ideas to achieve this goal: the dependency graph and the approximation of the ODE due to the cumulative propensity.

The dependency graph [5] is widely deployed to improve the performance of stochastic simulation by recording for each reaction the set of related reactions that may be affected when this reaction occurs. This idea can also help in the case of hybrid simulation by deciding, each time a stochastic reaction occurs, whether the ODE solver should be reinitialised. If there is any dependency between the just occurred reaction and any of the continuous reactions, the ODE solver should be reinitialised.

Furthermore, we isolate the ODE due to the cumulative propensities so that the deterministic solver becomes independent from the stochastic part for longer time periods. We replace the simultaneous numerical integration by an approximation of this ODE through one Euler step. This approach works well when there are only a few interface nodes between the stochastic and the deterministic regime, but a significant number of stochastic events. Fortunately, these two conditions are often satisfied for many real and complex biological models.

This paper is organised as follows: we start with an overview of biological network simulation by summarising the three main simulation approaches. Next, the proposed method of improving the hybrid simulation is presented by introducing two algorithms: one algorithm for finding the interface reactions, and the other one makes use of this information to improve the hybrid simulation. To illustrate the performance as well as the accuracy of this method, we show the result of three numerical experiments. Finally, we conclude with a few remarks and an outlook on future work.

2 Simulation of Biochemical Reaction Networks

We consider the problem of simulating N biochemical species $S_1, S_2, \ldots, S_N$ interacting through M reaction channels $r_1, r_2, \ldots, r_M$ in a well-mixed system of molecules [8]. Each reaction is associated with a rate constant k_i which serves as parameter for the state-dependent reaction rate. The state-dependent reaction rates usually follow specific kinetic rate laws, e.g., mass action, as all examples used in this paper.

There are many different simulation methods; see [8,26] for surveys. However, all of those procedures can be classified into three main approaches: deterministic, stochastic, or hybrid.

The traditional approach for studying the evolution of individual species with respect to time is to consider the chemical species evolving deterministically and continuously over time. According to this view, an ordinary differential equation (ODE) is constructed for each species S_i, and each reaction, in which S_i takes part, contributes to the species' ODE. The generation as well as the numerical solution of the system of ODEs are straightforward, but the simulation results do not always offer an accurate approximation of the dynamics of the corresponding biochemical system [8,25,26,30]. For instance, when one or more of the species participating in the biochemical reaction system exhibit a few number of molecules, the ODE approach fails to account for intrinsic noise due to the molecular fluctuation of such species [8,30]. Unfortunately, such fluctuations are often crucial when simulating biological systems; see, e.g., [2,22,30].

As an alternative approach, a stochastic simulation can be used by solving the Chemical Master Equation (CME), which calculates the probability that each species will have a certain number of molecules at a future time [8], or following the idea of Gillespie [6] by generating a set of trajectories by means of the stochastic simulation algorithm (SSA). In fact, the SSA algorithm generates a numerical realisations of the CME [8]. There are different variations of Gillespie's basic idea as well as many improvements [5, 26].

Nevertheless, the stochastic simulation is prohibitively slow to simulate real biological models that include many reaction events. As a discrete simulation, the stochastic simulation executes each individual event. For instance, in [21] the authors were not able to experiment with stochastic simulation due to the involvement of species with a big abundance of molecules. The performance of the stochastic simulation generally depends on the number of molecules of the participating species. For instance, for certain models, if the initial values of certain species are doubled, the time required to simulate the model rapidly increases [11].

With the recent increasing interest in using stochastic simulation to study the behaviour of biological models, new methods have been developed to improve the performance of the stochastic simulation by approximating the dynamics of biological systems. For example, in [4, 27] approximation techniques of the exact stochastic simulation have been developed by combining a number of events and execute them together instead of just simulating one reaction at a time. This improves the performance of the stochastic simulation by taking longer time steps. However, one difficulty of such a method is the selection of a suitable time step that makes a good trade-off between the simulation accuracy and efficiency.

Another direction to overcome the limitation of stochastic simulation is to resort to hybrid simulation [11, 23]. In this approach, the set of reactions are partitioned (either statically or dynamically) into two groups: slow and fast. The slow group contains reactions that have rate constants with small values and their corresponding substrate species do not exhibit high number of molecules, which yields in total low reaction rates, while the fast group contains the set of reactions with large rate constants and/or substrate species with abundant number of molecules. Afterwards, the slow group is simulated stochastically, while the fast group is simulated continuously.

The reactions in the two groups need to be synchronised due to interfacing species and/or reactions. With other words, the hybrid simulation algorithm has to decide when to jump from one simulation regime to the other. This step is crucial for the performance of hybrid simulation, and different synchronisation mechanisms have been proposed [11, 23, 26]. This synchronisation procedure is occasionally referred to as time transformation, since it switches the simulation time calculation from the continuous to the stochastic one.

Gillespie derives in [7] a general method for time-dependent propensities. This equation has been used afterwards in [1, 9, 11] to decide when to jump from the deterministic simulator to the stochastic one. It calculates the time when a stochastic event is to occur, while the ODE solver is integrating the system of

ODEs. According to the time-dependent propensity, a stochastic event occurs when the following equation is satisfied [11]:

$$\int_t^{t+\tau} a_0^s(\mathbf{x})dt + log(p_1) = 0, \tag{1}$$

where $\mathbf{x}$ is the state vector of the model at time t, p_1 is a random number uniformly distributed in $[0, 1]$, and a_0^s is the cumulative (total) propensity of the reactions in the slow group.

Equation (1) means that we carry out the numerical integration of the cumulative propensities that correspond to the slow reactions from the occurrence time of the previous reaction until the sum of the integral value and the log of a random number p_1 is equal to zero.

After that, the next reaction to occur is selected as the first index μ satisfying

$$\sum_{j=1}^{\mu} a_j^s(\mathbf{x}) > p_2 a_0^s(\mathbf{x}), \tag{2}$$

where p_2 is a random number generated from the uniform distribution, and a_j^s is the propensity of the jth slow reaction.

In [1] a set of algorithms have been derived to move from the deterministic regime to the stochastic one, when (1) is satisfied. The authors suggest to use an ODE solver with event handling to accurately detect the exact time point when (1) is satisfied. However, in [11] it has been previously asserted that exactly satisfying (1) will rapidly slow down the simulation. Therefore, in [11] they use an approximation of Eq. (1) to generate the trajectories by introducing a probability of no reaction (a dummy reaction).

However, this approximation approach will miss important events which has a negative impact on the accuracy of the hybrid simulator. Using an exact detection of (1), where the time-varying propensity is fully captured, will result in a hybrid simulator which is even slower than the pure stochastic one. This happens when many stochastic events occur in the slow part. This drawback has been practically encountered and reported by many researchers (see e.g., [24]).

One reason for this limitation is the frequent switch to the ODE solver. When there are many events in the stochastic regime, the cumulative propensity of slow reactions becomes larger and time steps between two successive stochastic events become smaller. Thus the time steps taken by the ODE solver will become smaller, too. So, the time saved by approximating some reactions using the deterministic simulation will be lost. Moreover, each time a stochastic event is fired, there is a discontinuity in the system of ODEs. Thus, the ODE solver should be reinitialised.

Frequent reinitialisation of the ODE solver decreases the performance of the whole hybrid simulation. Fortunately, not all firing events directly affect the system state of the ODE solver. Only reactions related to the interfacing species have an effect which needs to be taken into account. When a stochastic event occurs, which is not related to any interface reaction, the deterministic solver can simply continue working, after the firing of the stochastic event. On the contrary, when the occurring event is related to an interface reaction, the deterministic solver should restart the integration from this time point on.

In the subsequent section, we propose an algorithm to identify those interfacing reactions and then we show how the hybrid simulation can take advantage of this information to accelerate the overall simulation.

3   Accelerated Hybrid Simulation

In this section, we discuss the proposed procedure to improve the performance of the hybrid simulation. We start with describing the process of identifying interface reactions in the slow group that have a direct connection to the set of reactions in the fast group. Next, we outline the steps of the hybrid simulation that makes use of the information collected by the interface reaction detection procedure.

3.1   Detecting Interface Reactions

Suppose we have two groups of reactions: slow and fast. The occurrence of a reaction in the slow group can directly or indirectly influence the state vector of the variables in the fast group that are numerically integrated by an ODE solver.

A *direct dependency* exists between the two groups when the occurrence of a stochastic event changes one or more values of the state vector related to the fast reactions. This happens when one or more variables are shared between the two groups.

On the contrary, an *indirect dependency* exists when the firing of a stochastic event does not change any variable values related to the fast reactions, but it alters certain variables in the slow group which have an impact on the rates of the variables in the fast group during the numerical integration. That is, discrete variables appear in this case as multiplication factor with the rate constant (e.g., as in reactions related to positive and negative feedback loops).

For an example, consider the following two reaction sets:

$$r_1 : \; S_1 + S_2 \xrightarrow{k_1} P_1$$
$$r_2 : \; P_1 + S_3 \xrightarrow{k_2} P_2 \tag{3}$$

$$r_3 : \; S_1 + S_2 \xrightarrow{k_1} P_1$$
$$r_4 : \; S_3 + S_4 \xrightarrow{k_2 * P_1} P_2 \tag{4}$$

For the two reactions in (3), r_1 and r_2 share a common species (P_1). Therefore, when r_1 takes place, it affects the system state significant for the reaction r_2. But, for the reaction set in (4), when r_3 occurs, it does not directly affect the state of the species participating in r_4, but it influences the kinetic rate of this reaction.

Now consider the case, where r_1 is simulated stochastically, while r_2 is simulated deterministically by solving an ODE. Each firing of r_1 results in a discontinuity in the ODE involving r_2. But, when r_3 fires, it does not always cause discontinuities; such changes can be easily taken into account during the evaluation of the ODE function. Assuming that r_1 and r_2 are part of a bigger reaction network, then we call r_1 an *interface reaction*. When we list all interface reactions, we consider only those ones that have a direct dependency with the fast group.

Please note, we assume that the ODE solver is able to deal with small changes due the effect of indirect dependency of reactions. If this is not the case, then such reactions should also be added to the set of interface reactions. For instance, if the reaction r_3 increases the number of molecules of P_1 from 0 to 1, an abrupt change occurs which causes a discontinuity in the reaction set (4). As this behaviour does not occur during the whole simulation time, the specific time step where r_3 is anticipated to cause a discontinuity can be located during the simulation and therefore the ODE solver is reinitialised.

To find the set of all interface reactions whose occurrences affect the state vector of the ODE solver, we use an idea similar to the dependency graph [5]. The dependency graph is popular in accelerating stochastic simulations by storing for each reaction the set of other reactions that will be affected by every occurrence, so that the propensities do not have to be updated for all reactions each time a given reaction takes place. Please note that the dependency graph is created only once, but is extensively used during the simulation. Only the propensities of these dependent reactions are updated. Therefore, a substantial amount of processing time is saved in general. We use this information to detect the dependency between each stochastic reaction in the slow group and other related reactions in the fast group. Unlike the dependency graph used in the stochastic simulation, we record only those reactions that have a direct dependency with the fast group. Algorithm 1 summarises the steps for detecting all interface reactions that have a direct influence on the fast group.

The algorithm takes as input the two reaction groups: G_s and G_f, which denote the sets of slow and fast reactions, respectively. The set of interface reactions R^* is initially empty (step 1). Next, in steps 2–10, the procedure iterates over all reactions in the slow group. For each reaction r_i we identify the set of species that are altered when this reaction takes place (step 3). We call such variables *manipulated species*. This information can easily be found by help of the state change vector corresponding to such a reaction. After that the algorithm iterates for each manipulated species to find other reactions in the fast group that also manipulate the same species (that is it adds or substracts molecules from that species, when the reaction occurs) (step 5). If such a reaction is found, we mark r_i as an interface reaction (steps 6–7). On termination the algorithm returns the set of marked interface reactions (step 11).

At step 5, the algorithm finds first the set of reactions that manipulate the species in the slow group, then it searches for the existence of these species in the fast group. While this step seems to be redundant, the algorithm assumes that this information already exists through the dependency of all reactions

---

Algorithm 1. Finding Interface Reactions

Require: G_s the set of slow reactions;
Require: G_f the set of fast reactions;
 1: $R^* = \phi$ {the set of marked interface reactions is initially empty}
 2: **for each** r_i in G_s **do**
 3:    let S_i denotes the set of manipulated species when r_i fires;
 4:    **for each** s_{ij} in S_i **do**
 5:       Find the set of other reactions, R_{ij}, that manipulate s_{ij} when they fire;
 6:       **if** $\exists r_j \in R_{ij}$ and $r_j \in G_f$ **then**
 7:          Add r_i to R^*; {Mark r_i as an interface reaction}
 8:       **end if**
 9:    **end for**
10: **end for**
11: **return** R^*;

---

with each others, as it has been discussed in [5]. This dependency information is also required to increase the performance of the stochastic simulation of the slow subnet when updating the propensities of a fired reaction along with their dependent ones. If this information is not available, then we can alternatively search the reactions that manipulate s_{ij} directly from the fast reaction set.

As an example for the steps performed by Algorithm 1, consider the reactions in Table 1, assuming that these reactions are partitioned into two groups: the first group contains the slow reactions $G_s = \{r_1, \ldots r_5\}$, while the second group contains the set of fast reactions $G_f = \{r_6, \ldots r_8\}$. The set of manipulated species for the reactions r_1, r_2, r_3, r_4 are $\{A\}, \{A\}, \{B\}, \{B\}$, respectively. None of them is manipulated by any of the reactions in the fast group. Similarly, the set of manipulated species of the reaction r_5 is $\{A, B, C\}$. A and B are not manipulated by any reaction in the fast group. However, C is manipulated by the two reactions r_6 and r_7. Therefore, the reaction r_5 is identified as an interface reaction.

Obviously, as the number of interface reactions increases, the number of times the ODE solver is initialised also increases. This means that it is always advantageous to minimise the elements in the set of interface reactions while performing the partitioning. In what follows, we show how the information collected by this algorithm helps to accelerate the hybrid simulation.

3.2   Improving the Performance of the Hybrid Simulation Algorithm

Algorithm 2 lists the proposed steps to speed up the hybrid simulation algorithms presented in [1,11] by introducing two additional improvements: exploiting the dependency information collected by Algorithm 1, and replacing the simultaneous integration of the system of ODEs and Eq. (1) by a fixed integration step.

Reinitialising the ODE solver each time an event occurs at the stochastic regime predominately affects the performance of the hybrid simulator. As soon as the number of stochastic events increases, the repetitive reinitialisations of the ODE solver consumes most of the time of the hybrid simulator to switch from

Table 1. An example of a set of slow and fast reactions

#	Slow reactions	#	Fast reactions
r_1	$\phi \xrightarrow{s} A$	r_6	$C + E \xrightarrow{k_2} D$
r_2	$A \xrightarrow{d} \phi$	r_7	$D \xrightarrow{k_3} C + E$
r_3	$\phi \xrightarrow{s} B$	r_8	$D \xrightarrow{dd} \phi$
r_4	$B \xrightarrow{d} \phi$		
r_5	$A + B \xrightarrow{k_1} B + C$		

the stochastic to the continuous regime. In fact, this step hampers the practical implementation of hybrid simulation due to the computational expense required to reinitialise the ODE solver.

To clarify this point, consider the task of selecting an appropriate ODE solver to numerically integrate a system of ODEs. One can choose between simple fixed step size solvers (e.g., Euler) [10] or a more complicated numerical adaptive integration algorithm (e.g., backward differentiation formula [20]). Fixed step size solvers do not provide a good accuracy unless a very small step is chosen. However, using small steps rapidly degreases the performance of the simulation algorithm. Besides, even when using a small step size, the result is not satisfactory for many real applications, and such type of solvers do not scale well with the currently observed rapid increase of the systems biology's model sizes.

As an alternative to fixed step size solvers, adaptive ODE solvers can be used. To increase the performance of the numerical integrator, adaptive ODE solvers take small steps when the solution is not smooth and longer ones when the solution is smooth. Initially, such type of solvers employ certain algorithms to select a step to start with. Moreover, many libraries implement more or less complicated procedures to decide which step size to take by conducting many accuracy and convergence checks.

For instance, in a typical ODE library (e.g., [20]), each time the solver takes a step, it uses its own history information to advance the solution. Such type of solvers are highly required to improve the performance of hybrid simulation. However, in hybrid simulation, each time a stochastic event occurs, we have to clear all of this information by reinitialising the ODE solver, so that it can deal with the discontinuity due to the firing of a reaction in the slow regime. Moreover, the ODE solver will have to start from scratch each time this information is cleared. Indeed many libraries offer a lightweight reinitialisation to decrease the effect of this problem. However, also this version of reinitialisation takes a considerable time.

When we reset the ODE solver just a few times as in simple discrete-continuous systems, this issue does not present an intricate problem, since only a few discrete events will take place. However, for hybrid simulation it is compulsory to repeatedly clear the solution history each time an event occurs. Unfortunately, even for simple hybrid models there are thousands of such events (cf. Table 2).

Therefore, reducing the number of times the ODE solver is reinitialised will inevitably improve the performance of hybrid simulation.

Nevertheless, the hybrid simulation approaches in [1,11] require the simultaneous integration of the cumulative propensity (Eq. (1)) with the rest of the system of ODEs. This condition mandates the reinitialisation of the ODE solver each time an event occurs so that we restart the integration of (1) each time the state vector of the slow group is changed. It has been asserted in [11] that this step is computationally very expensive.

To relax this condition, we approximate the simultaneous integration of (1) by calculating its value using

$$a_0^s(\mathbf{x}) \cdot \Delta\tau + log(p_1) = 0, \tag{5}$$

where $\Delta\tau$ is the time difference between the occurrence time of the previous event and the current event, and p_1 is a random number generated from the uniform distribution.

Therefore, each time the ODE solver checks for the fulfillment of (1), it computes the value of (5) instead. The latter equation does not require the reinitialisation of the ODE solver, because there is no additional variable added to the system of ODEs that represents the cumulative propensity. When there are a substantial number of stochastic events, $\Delta\tau$ becomes very small which results in a good approximation of (1).

Algorithm 2 summarises the steps involved in improving the performance of the hybrid simulation algorithm in [11]. This algorithm takes advantage of the dependency graph in two locations. First, when it returns from the continuous integration, it updates all the propensities of stochastic reactions that share a species with the fast group. Second, it utilises the dependency information to decide whether to clear the ODE solver history information. We define two functions for this purpose: **Base**(r_i) and **Manipulated**(r_i). **Base**(r_i) returns all the species that are used to define the propensity of the reaction r_i, while **Manipulated**(r_i) returns all species that are affected by the occurrence of the reaction r_i. This information can easily be calculated from the dependancy graph.

The algorithm takes as input the two sets of reactions G_s and G_f as well as the set of marked interface reactions (R^*). At step 1, we initialise the ODE solver with the current concentration of species in the fast group. Initially, at step 2, we set the current time (τ) as well as the previous event time (τ_{old}) to zero. Afterwards the algorithm iterates until the end simulation time is reached (steps 3–16). At each iteration two random numbers $(p_1$ and $p_2)$ are generated (step 3). p_1 is used to calculate (5), while p_2 is used to find the the next event in (2). In the steps 5–7, we repeatedly integrate the system of ODE until (5) is satisfied. Equation (5) is recalculated each time the ODE solver checks for a root. When (5) is satisfied (step 7), the algorithm exits from the numerical integration to find the next stochastic event to occur using (2) (step 9). However, before finding the next stochastic reaction to fire, at step 8, the propensities of the reactions in the slow group are updated first using the information collected by the dependency graph. The function **Update**$(a(r_i), a_0^s)$ updates the propensity

---

Algorithm 2. Accelerated Hybrid Simulation

Require: G_s and G_f: the sets of slow and fast reactions respectively;
Require: R^* the set of reactions marked as interface reactions;
  1: Initialise the ODE solver with the initial concentration of the variables in G_f;
  2: set $\tau = \tau_{old} = 0$;
  3: **while** we did not reach end simulation time **do**
  4:     Generate two random numbers p_1 and p_2 from the uniform distribution;
  5:     **repeat**
  6:         Numerically integrate the system of ODEs;
  7:     **until** $a_0^s(\mathbf{x}) \cdot (\tau - \tau_{old}) + log(p_1) = 0$ {cf., Eq., (5)}
  8:     **Update**$(a(r_i), a_0^s)$, $\forall r_i \in G_s, \forall r_j \in G_f : $**Base**$(r_i) \cap $**Manipulated**$(r_j) \neq \phi$;
  9:     Find the reaction r_μ that satisfies (2) using p_2;
 10:     Fire r_μ and update the system state as well as the current time τ;
 11:     **Update**$(a(r_i), a_0^s)$, $\forall r_i : $**Base**$(r_i) \cap $**Manipulated**$(r_\mu) \neq \phi$;
 12:     Set $\tau_{old} = \tau$
 13:     **if** $r_\mu \in R^*$ **then**
 14:         Reinitialise the ODE solver
 15:     **end if**
 16: **end while**

---

of the reaction r_i ($a(r_i)$) as well as the cumulative one (a_0^s). Afterwards, the selected reaction is fired and the system state as well as the reaction propensities are updated (steps 9–11). The current event time is recorded as the previous event occurrence time (step 12). Thereafter the algorithm checks if the occurred reaction is in the marked list. If so, the ODE solver is reinitialised (steps (13–15)).

There are two updates of the slow reaction propensities. The first update takes place when we switch from the deterministic to the stochastic regime, while the second update is required when a reaction in the slow set has fired. In the former, all slow reactions that have a dependency with the fast reaction are updated, while in the latter, only reactions that have a dependency with the fired reaction are updated.

Algorithm 2 follows the direct method to implement hybrid simulation. However, it can be extended to include the first and next reaction methods as it has been proposed in [1,11,28]. But in our opinion, the direct method is the best choice in terms of implementation and performance unless a parallelisation approach is used. For instance, the first reaction method requires the generation of random numbers equal to the number of reactions in the reaction network which turns out to be a performance bottleneck. This drawback can be tackled by the next reaction method, but the latter complicates the implementation by introducing a specialised data structure: the priority queue.

Furthermore, this procedure assumes that the set of reactions are partitioned offline into slow and fast ones. However, this assumption does not prevent the use of an adaptive scheme as the one presented in [9] to run this algorithm. Nevertheless, the set of interface reactions will need to be updated each time a repartitioning is performed. To minimise the computational overhead due to the re-computation of the set of interface reactions, the already found interface reactions can be updated (e.g., by adding/removing reactions) instead of searching for them from scratch. Moreover, minimising the number of interface reactions can be introduced as an additional criteria for the dynamic partitioning procedure.

The check in step 13 can decrease the performance of the simulation algorithm if it will be implemented by searching the list of interface reactions each time a slow reaction is fired. However, a frequent search for the interface reactions can be avoided by storing a flag for each reaction that determines whether this reaction is marked as an interface one.

In the next section we show some experimental results of this algorithm.

4 Numerical Experiments

In this section, we present three examples to assess the performance of our accelerated hybrid simulation approach: a model for the circadian oscillation and two versions of a cell cycle model. In all examples, molecular fluctuations play a crucial role such that stochastic simulation is mandatory to reproduce the corresponding actual biological behaviours. We compare the performance of the exact algorithm with the accelerated one. For this purpose, we use the number of stochastic events in the slow group as a measure of the closeness of the simulation results.

These case studies have been tested using an implementation of Algorithms 1 and 2 in Snoopy [12] – a Petri net editing and simulation tool, and its related Steering Server for Collaborative Simulation, S^4 [15]. This implementation allows the simulation of (coloured) hybrid Petri nets [14,16]. An external ODE solver from the Sundials CVODE library [20] is used to simulate the deterministic part, while the direct method [6] is used to simulate the stochastic part. The runtime as well as the number of stochastic events are determined by executing the hybrid simulation 10 times and taking the average.

We provide all details for the first case study, which immediately permit to reproduce our results reported, but refer to the literature for the other two case studies because of the large number of reactions and species involved.

4.1 Circadian Oscillation

We use a rather small example with a few reactions and species to initially test the performance of the accelerated hybrid simulation. We adopt the circadian oscillation model introduced in [32]. The set of reactions as well as the kinetic rate constants are given in Appendix A.

The model comprises two genes: *gene1* and *gene2*. The two genes are transcribed into their corresponding mRNAs, which in turn are translated into proteins A (activator) and R (repressor), respectively. The reactions for the two genes/mRNAs/proteins are similar to each other, unless for those reactions that model the interaction between the two proteins.

This simple circuit together with the corresponding kinetic information fails to produce any oscillation when deterministic simulation is used [32]. Therefore,

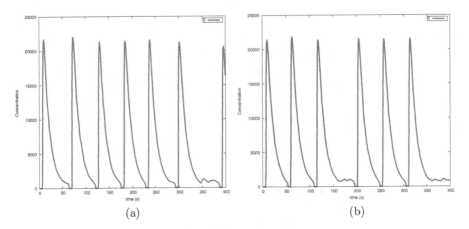

Fig. 1. Simulation results of the circadian oscillation model in [32] (single run) for: (a) exact, and (b) accelerated hybrid simulation

stochastic and hybrid simulation play a crucial role to simulate this model. We partition this reaction network such that reactions related to the mRNA of gene2 (R_{11}, R_{12}, R_{13}) are simulated stochastically, while the remaining reactions are executed via the deterministic solver. Such type of partitioning is sufficient to force the model to produce oscillations. Fortunately, the set of marked reactions (R^*) is empty since the three slow reactions do not influence the state vector of the fast reactions.

Figure 1 depicts the simulation result of this model when it has been executed using both the exact as well as the accelerated version of the hybrid simulation algorithm. Moreover, Table 2 lists the runtime behaviours of the two algorithms.

According to the data in Table 2, there is an improvement in terms of runtime when the accelerated algorithm is used. The accelerated simulation algorithm is about three times faster than the exact method. Moreover, the number of generated stochastic events for both simulators are comparable. Therefore, this saving in runtime behaviour is due to the avoidance of repeated reinitialisation of the ODE solver and not because of reducing the number of stochastic events.

4.2 Yeast Cell Cycle

As another test and motivation example for the improvements presented in this paper, we deploy cell cycle modelling [31]. Cell cycle models are used to study the regulation of a cell during its replication and division. There are many kinetic models in the literature to study such biological behaviour. One important aspect of these models is that it is crucial to capture intrinsic noise in order to reproduce wet-lab experiments related to the variation in the age and volume of daughter cells [2,22,31]. In [22], Kar et al., constructed a stochastic model to study the effects of molecular fluctuation of species with low copies of molecules on the variation of cellular volume. This model includes many species with different

Table 2. Statistical information related to the three case studies used to evaluate the performance of Algorithm 2

Criteria/models	Circadian Oscillation	Cell Cycle Model (V1)	Cell Cycle Model (V2)
Number of species	9	26	60
Number of reactions	16	48	190
Number of stochastic reactions	3	19	19
Number of deterministic reactions	13	29	171
Number of stochastic events (exact)	35,650	780,318	112,908
Number of stochastic events (accelerated)	35,533	776,192	112,789
Run times (exact) (s)	3.8	731	495
Run times (accelerated) (s)	1.278	445	53
Number of interface reactions	0	8	0

abundance of molecules. Therefore, in [17, 19, 24], it has been suggested to use hybrid simulation to study the dynamics of this model. However, it has also been reported there that the hybrid simulation algorithm is computationally very slow when simulating this model [24].

In [2], a new deterministic and stochastic model to study the yeast cell cycle has been developed. The dimension of the model has been substantially increased, as it is based on the approach of multi-site phosphorylation to reproduce the level of nonlinearity that is required to reconstruct the bistable switch behaviour. Similarly, stochastic simulation is of paramount importance to capture the variation of the cellular volume. However, simulation becomes more intricate due to the many species and reactions involved in the construction of this model.

In this section we use the accelerated hybrid simulation algorithm to study the behaviour of this model. We adopt two model versions. The simpler one (V1) is based on the stochastic and hybrid models constructed in [22] and [19], respectively, and the bigger one (V2) on the model in [2]. A summary of the model information as well as the result statistics of applying Algorithm 2 are given in Table 2. To run the simulation, we use the same kinetic information as well as initial state as given in [2, 22].

The two models have been statically partitioned such that all reactions related to mRNAs are stochastically simulated, while all other reactions are deterministically simulated. Figures 2 and 3 present the simulation results of a single run of the exact as well as the accelerated version of the hybrid simulation for the cell cycle model V1 and V2, respectively. Both simulator versions successfully capture the variation of the cell volume over the whole simulation time. However, the accelerated simulator version is ten times faster than the exact one as it has been illustrated in Table 2.

We use again the number of stochastic events as a measure of the simulation accuracy. According to the data in Table 2, both simulation approaches generate

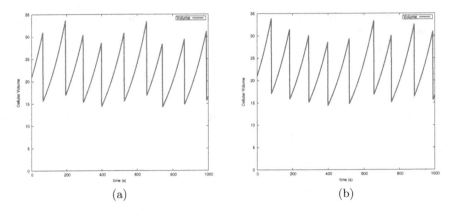

Fig. 2. Results of the yeast cell cycle model V1 (cellular volume) when simulated via (a) the exact algorithm, (b) the accelerated algorithm. Please note that single runs were used to generate this figure

comparable numbers of stochastic events. However, as the many reinitialisations of the ODE solver have been avoided, the accelerated version runs much faster than the exact one. As another measure of simulation accuracy, we refer to the number of divisions during the whole simulation period. The cell cycle model V1 produces nine divisions in both the exact and accelerated simulators, while the cell cycle model V2 results in five divisions when running the two simulators.

The cell cycle model in its first version (V1) is much smaller in terms of number of species and reactions compared to the larger second version (V2). However, the number of stochastic events is much higher due to the abundance

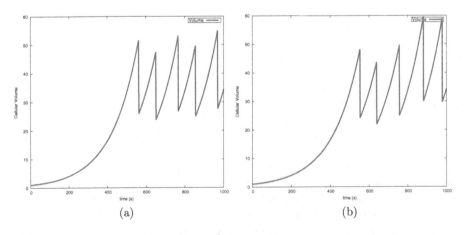

Fig. 3. Results of the yeast cell cycle model V2 (cellular volume) when simulated via (a) the exact algorithm, (b) the accelerated algorithm. Please note that single runs were used to generate this figure

of molecules of the mRNA species. Moreover, it was not feasible to completely isolate the deterministic and stochastic parts as we did for V2. The cell cycle model V1 contains 8 interface reactions which have an noticeable effect on the runtime behaviour of the model simulation due to the required reinitialisations, while there are no interface reactions in V2.

5   Conclusions and Future Work

In this paper, we have presented an approach for improving the performance of hybrid simulation algorithms that combine stochastic and (continuous) deterministic solvers. The proposed method has been tested using three case studies; and the test results have also been presented. The suggested improvements will be useful to cope with the rapid growth of (biological) models.

According to the test results given in Table 2 it is obvious that there is a noticeable improvement in terms of runtime even for small models. The speed of the accelerated algorithm is about three times faster than the exact method. However, this is not very significant since there are only a few continuous variables used inside the ODE solver. Avoiding the reinitialisation in this case will not save much runtime.

On the contrary, for bigger models, there is a substantial improvement in terms of runtime (about ten times faster). This result is due to saving the ODE solver from repeating the work required to build accuracy and history information. Therefore, as soon as the model size is increased, the simulator performance is also improved. Moreover, there is another advantage when using these improvements for larger models: as the number of stochastic events increases, the time step between two successive events will decrease. This will also increase the algorithm accuracy due to the calculation of the cumulative propensity via Eq. (5).

For smaller models with a few stochastic events, the time steps between two successive events will be larger. Therefore, the calculation in Eq. (5) will decrease the simulation accuracy. One workaround to this limitation is to switch to the exact method whenever the time step is large, while using the accelerated method when the time step is small. One measure to help deciding (dynamically) which approach to use is the current value of the cumulative propensity.

The partitioning process including the dependency algorithm presented in this paper can be automated such that users of the hybrid simulator obtain suggestions for the best partitioning settings that result in an optimal performance of the hybrid simulation, while pertaining an acceptable level of accuracy.

Furthermore, the set of slow and fast reactions are partitioned so that there is no direct dependency between the two groups. Fortunately, this has completely eliminated the reinitialisation step for each switch from the stochastic to the continuous solvers. While completely isolated subnets are not common for all biological model (cf., cell cycle model V1), the partitioning process can be designed so that the set of interface reactions is minimal.

Finally, in Sect. 4 we have presented three case studies to assess the performance of the proposed method. Adopting more examples with different sizes and partitioning strategies will provide more insights about the performance of Algorithm 2.

Acknowledgments. This work has been partially funded by the GE-SEED grant (7934) which is administrated by STDF(Science and Technology Development Fund, Egypt) and DAAD (German Academic Exchange Service). We also acknowledge the helpful comments of the anonymous reviewers for improving a previous version of the paper.

A   Reactions of the Circadian Oscillation Model

Table 3 provides a complete specification of the circadian oscillation model [32] used in Sect. 4.1, This reaction set has been derived from the system of ODEs given in [32] following the approach described in [13]. The given ODEs fulfil the criteria established in [29]; so the result is unique. See [3] for a graphical representation of this reaction set by use of Petri nets and their treatment in the different paradigms.

Table 3. The reaction set of the circadian oscillation model

#	Reaction	Propensity	Parameter Value
r_1	$gene1\_active \xrightarrow{k_1} gene1 + A$	$k_1 \cdot gene1\_active$	$k_1 = 50$
r_2	$gene1 + A \xrightarrow{k_2} gene1\_active$	$k_2 \cdot gene1 \cdot A$	$k_2 = 1.0$
r_3	$gene2\_active \xrightarrow{k_3} gene2 + A$	$k_3 \cdot gene2\_active$	$k_3 = 100$
r_4	$gene2 + A \xrightarrow{k_4} gene2\_active$	$k_4 \cdot gene2 \cdot A$	$k_4 = 1.0$
r_5	$\phi \xrightarrow{k_5} mRNA1$	$k_5 \cdot gene1\_active$	$k_5 = 500$
r_6	$\phi \xrightarrow{k_6} mRNA1$	$k_6 \cdot gene1$	$k_6 = 50$
r_7	$mRNA1 \xrightarrow{k_7} \phi$	$k_7 \cdot mRNA1$	$k_7 = 10$
r_8	$\phi \xrightarrow{k_8} A$	$k_8 \cdot mRNA1$	$k_8 = 50$
r_9	$A \xrightarrow{k_9} \phi$	$k_9 \cdot A$	$k_9 = 1.0$
r_{10}	$A + R \xrightarrow{k_{10}} A\_R$	$k_{10} \cdot A \cdot R$	$k_{10} = 2.0$
r_{11}	$\phi \xrightarrow{k_{11}} mRNA2$	$k_{11} \cdot gene2\_active$	$k_{11} = 50$
r_{12}	$\phi \xrightarrow{k_{12}} mRNA2$	$k_{12} \cdot gene2$	$k_{12} = 0.01$
r_{13}	$mRNA2 \xrightarrow{k_{13}} \phi$	$k_{13} \cdot mRNA2$	$k_{13} = 0.5$
r_{14}	$\phi \xrightarrow{k_{14}} R$	$k_{14} \cdot mRNA2$	$k_{14} = 5$
r_{15}	$R \xrightarrow{k_{15}} \phi$	$k_{15} \cdot R$	$k_{15} = 0.08$
r_{16}	$A\_R \xrightarrow{k_{16}} R$	$k_{16} \cdot A\_R$	$k_{16} = 1.0$

References

1. Alfonsi, A., Cancès, E., Turinici, G., Ventura, B., Huisinga, W.: Adaptive simulation of hybrid stochastic and deterministic models for biochemical systems. ESAIM: Proc. **14**, 1–13 (2005)
2. Barik, D., Baumann, W.T., Paul, M.R., Novak, B., Tyson, J.J.: A model of yeast cell-cycle regulation based on multisite phosphorylation. Molecular Syst. Biol. **6**(1), 405 (2010)
3. Blätke, M., Heiner, M., Marwan, W.: BioModel engineering with Petri nets, chap. 7, pp. 141–193. Elsevier Inc. (2015). http://store.elsevier.com/product.jsp?isbn=9780128012130
4. Cao, Y., Gillespie, D., Petzold, L.: Adaptive explicit-implicit tau-leaping method with automatic tau selection. J. Chem. Phys **126**(22), 224101 (2007)
5. Gibson, M., Bruck, J.: Exact stochastic simulation of chemical systems with many species and many channels. J. Phys. Chem. **105**, 1876–89 (2000)
6. Gillespie, D.: Exact stochastic simulation of coupled chemical reactions. J. Phys. Chem. **81**(25), 2340–2361 (1977)
7. Gillespie, D.: Markov Processes: An Introduction for Physical Scientists. Academic Press, San Diego (1991)
8. Gillespie, D.: Stochastic simulation of chemical kinetics. Annual Rev. Phys. Chem. **58**(1), 35–55 (2007)
9. Griffith, M., Courtney, T., Peccoud, J., Sanders, W.H.: Dynamic partitioning for hybrid simulation of the bistable HIV-1 transactivation network. Bioinformatics **22**(22), 2782–2789 (2006)
10. Hairer, E., Wanner, G.: Solving Ordinary Differential Equations II: Stiff and Differential-Algebraic Problems, Springer Series in Computer Mathematics. Springer Series in Computer Mathematics, vol. 14. Springer, Berlin (1996)
11. Haseltine, E., Rawlings, J.: Approximate simulation of coupled fast and slow reactions for stochastic chemical kinetics. J. Chem. Phys. **117**(15), 6959–6969 (2002)
12. Heiner, M., Herajy, M., Liu, F., Rohr, C., Schwarick, M.: Snoopy – a unifying Petri net tool. In: Haddad, S., Pomello, L. (eds.) PETRI NETS 2012. LNCS, vol. 7347, pp. 398–407. Springer, Heidelberg (2012)
13. Hellander, A., Lötstedt, P.: Hybrid method for the chemical master equation. J. Comput. Phys. **227**(1), 100–122 (2007)
14. Herajy, M., Heiner, M.: Hybrid representation and simulation of stiff biochemical networks. J. Nonlinear Anal. Hybrid Syst. **6**(4), 942–959 (2012)
15. Herajy, M., Heiner, M.: A steering server for collaborative simulation of quantitative Petri nets. In: Ciardo, G., Kindler, E. (eds.) PETRI NETS 2014. LNCS, vol. 8489, pp. 374–384. Springer, Heidelberg (2014)
16. Herajy, M., Liu, F., Rohr, C.: Coloured hybrid Petri nets for systems biology. In: Proceedings of the 5th International Workshop on Biological Processes & Petri Nets (BioPPN), Satellite Event of PETRI NETS 2014, CEUR Workshop Proceedings, vol. 1159, pp. 60–76 (2014)
17. Herajy, M., Schwarick, M.: A hybrid Petri net model of the eukaryotic cell cycle. In: Proceedings of the 3rd International Workshop on Biological Processes and Petri Nets (BioPPN), Satellite Event of PETRI NETS 2012, CEUR Workshop Proceedings, vol. 852, pp. 29–43 (2012). CEUR-WS.org. http://ceur-ws.org/Vol-852/
18. Herajy, M., Heiner, M.: Modeling and simulation of multi-scale environmental systems with generalized hybrid Petri nets. Front. Environ. Sci. **3**(53) (2015)

19. Herajy, M., Schwarick, M., Heiner, M.: Hybrid Petri nets for modelling the eukaryotic cell cycle. In: Koutny, M., Aalst, W.M.P., Yakovlev, A. (eds.) ToPNoC VIII. LNCS, vol. 8100, pp. 123–141. Springer, Heidelberg (2013). doi:10.1007/978-3-642-40465-8_7

20. Hindmarsh, A., Brown, P., Grant, K., Lee, S., Serban, R., Shumaker, D., Woodward, C.: Sundials: suite of nonlinear and differential/algebraic equation solvers. ACM Trans. Math. Softw. **31**, 363–396 (2005)

21. Iwamoto, K., Hamada, H., Eguchi, Y., Okamoto, M.: Stochasticity of intranuclear biochemical reaction processes controls the final decision of cell fate associated with DNA damage. PLoS ONE **9**, 1–12 (2014)

22. Kar, S., Baumann, W.T., Paul, M.R., Tyson, J.J.: Exploring the roles of noise in the eukaryotic cell cycle. Proc. Natl. Acad. Sci. U.S.A. **106**(16), 6471–6476 (2009)

23. Kiehl, T., Mattheyses, R., Simmons, M.: Hybrid simulation of cellular behavior. Bioinformatics **20**, 316–322 (2004)

24. Liu, Z., Pu, Y., Li, F., Shaffer, C.A., Hoops, S., Tyson, J.J., Cao, Y.: Hybrid modeling and simulation of stochastic effects on progression through the eukaryotic cell cycle. J. Chem. Phys. **136**(3), 034105 (2012)

25. Mcadams, H., Arkin, A.: It's a noisy business!. Trends Genet. **15**(2), 65–69 (1999)

26. Pahle, J.: Biochemical simulations: stochastic, approximate stochastic and hybrid approaches. Brief Bioinform. **10**(1), 53–64 (2009)

27. Rathinam, M., Petzold, L., Cao, Y., Gillespie, D.: Stiffness in stochastic chemically reacting systems: the implicit tau-leaping method. J. Chem. Phys. **119**, 12784–12794 (2003)

28. Salis, H., Kaznessis, Y.: Accurate hybrid stochastic simulation of a system of coupled chemical or biochemical reactions. J. Chem. Phys **122**(5), 54103 (2005)

29. Soliman, S., Heiner, M.: A unique transformation from ordinary differential equations to reaction networks. PLoS ONE **5**(12), e14284 (2010)

30. Srivastava, R., You, L., Summers, J., Yin, J.: Stochastic vs. deterministic modeling of intracellular viral kinetics. J. Theor. Biol. **218**(3), 309–321 (2002)

31. Tyson, J.J., Novk, B.: Irreversible transitions, bistability and checkpoint controls in the eukaryotic cell cycle: a systems-level understanding, Chapt. 14. In: Walhout, A.M., Vidal, M., Dekker, J. (eds.) Handbook of Systems Biology, pp. 265–285. Academic Press, San Diego (2013)

32. Vilar, J., Kueh, H., Barkai, N., Leibler, S.: Mechanisms of noise resistance in genetic oscillators. PNAS **99**, 59885992 (2002)

Hybrid Stochastic Simulation of Rule-Based Polymerization Models

Thilo Krüger[✉] and Verena Wolf

Modeling and Simulation Group, Saarland University, Saarbrücken, Germany
{thilo.krueger,verena.wolf}@uni-saarland.de

Abstract. Modeling and simulation of polymer formation is an important field of research not only in the material sciences but also in the life sciences due to the prominent role of processes such as actin filament formation and multivalent ligand-receptor interactions. While the advantages of a rule-based description of polymerizations has been successfully demonstrated, no efficient simulation of these mostly stiff processes is currently available, in particular for large system sizes.

We present a hybrid stochastic simulation approach, in which the average changes of highly abundant species due to fast reactions are deterministically simulated while for the remaining species with small counts a rule-based simulation is performed. We propose a nesting of rejection steps to arrive at an approach that is efficient and accurate. We test our method on two case studies of polymerization.

Keywords: Polymerization · Rule-based modeling · Hybrid simulation

1 Introduction

Polymerizations are chemical reaction processes, where a number of monomer molecules form polymer chains or more complex three-dimensional network structures. They play an important role in the life sciences since polymerizations naturally occur in living organisms. Prominent examples of biochemical polymerization processes are actin polymerization [24], multivalent ligand-receptor interactions [15,22] and plaque-formation during Alzheimer's disease [3,6,16]. In the material sciences polymerizations are considered in the context of synthetic polymers such as plastics and elastomers and simulations of polymerizations are used for performance optimization. Current research in this area is mostly focusing on deterministic models [19,21,30] but stochastic approaches based on Gillespie's simulation algorithm are also common [1,27].

Standard deterministic approaches, such as an equation system with one ordinary differential equation per chemical species, are usually not feasible due to the very large or infinite number of involved species, though some approximations have shown accurate results for certain variables of interest like the distribution of the chain length [29,30].

© Springer International Publishing AG 2016
E. Cinquemani and A. Donzé (Eds.): HSB 2016, LNBI 9957, pp. 39–53, 2016.
DOI: 10.1007/978-3-319-47151-8_3

Polymerizations have also been described by rule-based models, which allow a network-free stochastic simulation of their evolution [25,31]. Rule-based models describe a system by assigning different but structurally similar reactions to a rule. When simulating such models network-free, instead of choosing a reaction that has to be simulated, a rule is chosen. Then the reactants are chosen probabilistically based on the current state of the system. The performance of network-free stochastic simulations only depends on the number of such rules and not anymore on the number of reactions. This yields an enormous speed-up for systems with many reactions that can be described by a few rules. The generated trajectories are stochastically identical to those of the standard Gillespie algorithm [13]. Popular simulation tools like *NFSim* [25] or *KaSim* [9] use agent-based simulations, during which each individual molecule is tracked instead of populations of molecules. According to a few rules with known propensities, the reacting monomers can be chosen directly, and the reaction can be performed or it is rejected, because the corresponding product of the reaction is not possible in reality. This requires huge amounts of memory, on the other hand, the access to the reacting patterns is fast and rejection occurs not too often. However, agent-based simulations of polymerizations on a larger scale are infeasible due to the huge amounts of memory needed. To make large-scale simulations possible, a version of agent-based simulation was proposed, where a part of the species in the system is represented as populations and only species with an expected low count are treated as agents [18].

For polymerizations such simulation methods are, however, still inefficient since they have to perform a large number of fast bindings of/reactions between highly abundant species (e.g. monomers). Thus, until now this stiffness problem of polymerizations has not been adequately addressed. However, reaction networks with largely varying propensities have been extensively studied in the context of the Gillespie algorithm [13] and led to a large number of multi-step approaches including tau-leaping (see [14] for an overview), quasi steady-state approaches [4,5], and hybrid simulation approaches [7,17,23]. Hybrid simulation approaches simulate the reactions, that evolve on a fast time scale, deterministically during time intervals determined by the random times between the reactions on the slow time scale. Thus, they track the (deterministically changing) average molecule numbers for highly abundant species given a stochastically selected evolution for the molecule numbers of species present in low copy numbers.

Here, we propose a similar approach to stiff rule-based models and in particular to polymerization reactions, which are intrinsically stiff because species present in low and high copy numbers occur at the same time in the system. The propensities in such systems usually differ by several orders of magnitude and highly abundant species anyway do not need an expensive stochastic description as their evolution is sufficiently accurately captured by the average changes of their counts.

Application of a hybrid approach to network-free simulation algorithms is not straightforward since for an efficient simulation of the rules, a rejection step is necessary [31]. Moreover, since the propensities of slowly changing species depend on the molecular counts of highly abundant species, these propensities are no longer constant in time, i.e. they are coupled with the ODE system of the highly abundant species. We therefore suggest a nested rejection step that is necessary because we choose propensities that dominate the time-varying propensities and that stay constant until the next slow reaction occurs. The correctness of our approach is based on a successive application of the well-known thinning theorem by Lewis and Shedler [20] and similar approaches have been used in other contexts [2,26] as well as for time-dependent rates in rule-based models [3]. Since each rule describes a set of reactions, it is not easily possible to differentiate between fast and slow reactions. We therefore dynamically split the species into those of high and those of low count based on a threshold. We then only integrate those reactions deterministically that do not involve any species of low count. All other reactions are simulated according to a stochastic network-free simulation algorithm.

A comparison of the proposed hybrid simulation method with simulations using *NFSim* shows that the suggested method performs better in particular when the reaction system is stiff.

In Sect. 2, we shortly recall the basics of polymerizations and simulations of rule-based models. We then present in detail our hybrid approach in Sect. 3 and show results for two case studies in Sect. 4.

2   Background

2.1   Modeling of Polymerizations

There are generally two different mechanisms that lead to polymers, step-growth polymerizations and chain polymerizations [19]. The most common approach to model both kinds of polymerizations is the deterministic approach, i.e. defining a system of ordinary differential equations (ODEs) for the population change of each of the different species (i.e. types of molecules). We illustrate the two mechanisms by means of two examples (see Fig. 1) and give the corresponding ODE system. To concentrate on the principal mechanism of the chain elongation, further reactions like chain break or deactivation reactions are omitted here but considered for the results in Sect. 4. In step-growth polymerizations (Fig. 1, left) each monomer contains at least two reaction sites x and $x - P_n - x$ describes a molecule with two sites and a polymer chain of length n in between. Each site can react with each other site. Therefore, each molecule in a reaction system can react with each other molecule in the system. Hereby, the counts of modules (i.e. subunits in a polymer chain or network structure) in the product is the sum of the counts of each of the educts and the ODEs contain infinite sums. Usually, bonds in such polymers will also break with a low reaction rate, we omit these backward reactions in the ODEs.

In chain-polymerizations (Fig. 1, right) the monomers have one inactive reaction site u that has to be activated (a), for example by an initiator I^*. In a chain-elongation reaction, the activated site is shifted to the newly formed end of the chain, the count of modules is growing by one in each step and the ODEs are much simpler.

Monomer: x–P_1–x

Monomer: M–u

$$x-P_n-x + x-P_m-x \xrightleftharpoons{k} x-P_{m+n}-x$$

$$M-u + I^* \xrightarrow{init} P_1-a + I$$

$$\frac{dP_n}{dt} = k \left(\sum_{m+l==n} P_m P_l - \sum_{m=1}^{\infty} P_m P_n \right)$$

$$P_n-a + M-u \xrightarrow{k} P_{n+1}-a$$

$$\frac{dP_n}{dt} = kM(P_{n-1} - P_n)$$

Fig. 1. Reaction schemes for two types of polymerizations and the ODEs for the growth reaction: step-growth polymerizations (left) and chain-polymerizations (right).

As shown in Fig. 1, the ODE systems are infinite and can include infinite sums in the case of a step-growth polymerization. This issue is related to the combinatorial explosion problem that occurs when simulating polymerizations. Let us consider a theoretical step-growth copolymerization of two different monomers, where both monomers have two reaction sites. The number of different species that can occur during such a polymerization process and contain N monomers in total is greater than 2^{N-1}. Setting up ODEs for each possible polymer chain by hand is no longer easily done as in Fig. 1. This becomes even more difficult if one monomer has a third reaction site such as in the trivalent ligand - bivalent receptor model, where a polymer network is built (see also Sect. 4.2). The combinatorial explosion is not a major problem in linear chain-homopolymerizations, since the number of occurring species just scales with the chain length. The main driver for our work are therefore more complex polymerizations like step-growth copolymerizations or network building polymerizations.

Polymerizations can also be described by a stochastic model, whose mean-field is the corresponding ODE solution, and simulated using Gillespie's algorithm [13]. Complex polymerizations can be simulated by generating newly occurring species on the fly, i.e. only if they really occur during the simulation. However, such simulations will be very slow if a large number of monomers are initially present in the system. In typical polymerizations each species can react with at least one monomer. Hence, the number of possible reactions will scale at least with the number of species in the system. This renders Gillespie simulation of such systems slow because in each step of a simulation all possible reactions have to be considered. One way to circumvent this problem is to describe polymerizations by rule-based models and to analyze such models by performing a network-free simulation.

2.2 Rule-Based Modeling

In rule-based modeling different but structurally similar reactions are combined into rules [8,11,25]. For example, an activation of a functional group occurs at the same rate independent of the chemical environment of the functional group, i.e. of other proteins belonging to the same complex molecule. Polymerizations show a similar behavior: they undergo elongation (or shortening) reactions with a rate independent of the chain length, which can be described by a single rule.

In the following we concentrate on those details of rule-based modeling that are necessary for the presentation of the proposed hybrid simulation method and do not define a specification language for rules as in [10].

Formally, a given finite set of rules $R = \{R_1, R_2 \ldots\}$ of the form $Z_1 \xrightarrow{c} Z_2$ (order one) or $Z_1 + Z_2 \xrightarrow{c} Z_3$ (order two)[1] describes the reactions that all species matching the educt patterns Z_1 (and Z_2) can undergo. A pattern is a specific chemical structure that can be a molecule, a part of a molecule or a part of a molecule bound to one or more undefined wildcards (denoted by $W_1, W_2, \ldots$). To each pattern Z_i we assign a set of chemical species whose corresponding molecules match Z_i, i.e. the chemical structure of the pattern is part of the molecule and all wildcards are substituted appropriately. Depending on the kind of a pattern, a molecule can match a pattern multiple times. If the pattern contains a site that shall be modified according to a rule, a molecule matches a pattern as many times, as there are according sites part of the molecule. If a pattern contains a bond that shall break in a reaction, a molecule matches the pattern as many times as an according bond is part of the molecule. Thus, for each pattern Z_i, we assign not only a set of species, but also for each species how often the corresponding molecule matches the pattern, i.e. the multiplicity of the pattern. The corresponding product pattern will replace the educt pattern(s) after a rule was applied.

By finding all molecules in the simulated system that match the educt patterns of a rule, the rule implicitly describes a set of chemical reactions $S = \{S_1, S_2, \ldots\}$ of the form

$$X_1 \xrightarrow{c} X_2 + \cdots \qquad \text{or} \qquad X_1 + X_2 \xrightarrow{c} X_3 + \cdots$$

where $X_1, X_2, \ldots$ are chemical species with species counts $x_1, x_2, \ldots$ and c is the stochastic reaction rate constant. The key idea of the simulation of rule-based models is, instead of finding a reaction to simulate, to find a rule to simulate, and to instantiate the rule in a second step with a reaction, by finding species that match the educt patterns of the rule. Let α_{R_i} be the propensity of the i-th rule and α_{S_j} the propensity of the j-th reaction. Since one rule describes several reactions,

$$\alpha_{R_i} = \sum_{\substack{S_j \text{ is described by } R_i}} \alpha_{S_j} \geq \sum_{\substack{S_j \text{ is possible} \\ \text{and described by } R_i}} \alpha_{S_j}. \qquad (1)$$

To simulate such models efficiently, rules describe more reactions than only those that really can happen. For example, an elongation rule for step-growth polymerizations can be written as

$$W_1 - x + W_2 - x \longrightarrow W_1 - X - X - W_2.$$

---

[1] Here, we consider only a single product pattern for rules, since reactions with two products can still be described by a single product pattern. In such a case, the pattern does not represent one connected chemical structure, but the structure misses a bond and can therefore be divided into two substructures.

Such a rule describes all elongation reactions according to Fig. 1, left, but it also describes usually forbidden ring formations if one molecule matches both educt patterns. In a network-free simulation, if such a rule is chosen and a forbidden reaction is selected, we perform a rejection step. Note that this is much more efficient than checking beforehand for all reactions whether they are possible or not.

1: Initialize time t, x_1, x_2, ... and educt pattern counts $z_1,z_2,...$;
2: **repeat**
3:   **repeat**
4:     compute $\alpha_{R_1},...,\alpha_{R_m}$ and $\alpha_R = \sum_i \alpha_{R_i}$;
5:     draw an exponentially distributed pseudo random number τ with rate α_R and update $t = t + \tau$;
6:     draw a pseudo random number $I \in \{1, 2, ...\}$ according to the discrete distribution $\frac{\alpha_{R_1}}{\alpha_R}, \frac{\alpha_{R_2}}{\alpha_R}$, ... , and choose the corresponding rule R_I;
7:     instantiate the chosen rule by probabilistically selecting one reacting pattern Z_1 out of all existing educt patterns of R_I. Repeat the same in case of a rule of order two for Z_2;
8:     check if the reaction S_i that is defined by the rule and the chosen patterns is possible;
9:   **until** S_i is possible
10:    apply R_I to the species/sites and update x_1, x_2, ... and $z_1,z_2,...$;
11: **until** stop-condition is reached

Algorithm 1. Basic algorithm for simulating rule-based models.

We present in Algorithm 1 the pseudo-code of a network-free simulation algorithm. We denote the counts of all educt patterns as $z_1, z_2, ...$. These counts are tracked during the simulation, since they are needed to compute the propensities α_{R_i} in Line 4 and to instantiate the chosen rule in Line 7. Computing the propensities of rules is done similar to computing propensities of reactions, it is just the product of all educt pattern counts and the stochastic rate constant. For the instantiation of the rules, also the multiplicity of the patterns for each species is multiplied. Therefore, it is necessary to loop over all species that are currently present in order to probabilistically choose the pattern. The check, if the instantiation is possible (Line 8) can rely on arbitrary constraints and if it is successful, the product of the reaction has to be determined and whether further copies of this product are already in the system. This is the most time-consuming step of the simulation, since a tree-isomorphism check has to be performed. Note that the correctness of the simulation algorithm relies on the thinning theorem since the acceptance probability corresponds to the quotient of the two sums in Eq. 1 (see also [9]).

3   Hybrid Simulation Algorithm

In this section we propose a stochastic simulation algorithm for rule-based models of polymerizations that is network-free and splits the set of species of the corresponding reaction network into continuous and discrete species. We use $X = \{X_1, \ldots\}$ and $Y = \{Y_1, \ldots, \}$ to denote the set of discrete and continuous species and x_i to denote the current count of the discrete species X_i. For the continuous species, y_i denotes the *average* count of Y_i.

We also split the set of all instantiations of a rule R_i (i.e. reactions that are described by R_i, including those that are not possible) into those, for which *all educts and all products are continuous species*, denoted by $C_i = \{c_{i1}, c_{i2}, \ldots\}$ (continuous reactions), and those that contain (at least one) discrete species, denoted by $D_i = \{d_{i1}, d_{i2}, \ldots\}$ (discrete reactions). The main idea behind our hybrid approach is to deterministically simulate the c_{ij}'s instead of simulating them stochastically. Thus, we integrate a system of ODEs for the y_i's until the next discrete reaction occurs in the network-free stochastic simulation. Each $c_{ij} \in C_i$ describes a reaction and its influence to the change of a population count appears in the system of ODEs for the continuous species. During the simulation, the state of a system at time t is described by the current sizes of the discrete and continuous populations. Here, it is important to only consider those discrete species that have a positive count because the total number of discrete species is typically extremely large or infinite. Hence, the current set of discrete and continuous species considered during the simulation changes dynamically, since certain discrete species may have count zero or may reach a threshold $K \gg 1$. If the count of a discrete species becomes greater than K, then the species becomes a continuous species. Note that even for $K = 1$, we usually have discrete reactions since there are species with current count zero. In our implementation we use as a default value $K = 200$, and if the count of a continuous species becomes significantly lower than K (i.e. $K/2 = 100$), then the species becomes a discrete species. In this way we avoid frequent switching between discrete and continuous representations. Moreover, around this value the performance of the method is best for the examples that we considered. We discuss this issue in detail in Sect. 4.

Since continuous species may be educts of discrete reactions and hence influence the propensities of these reactions, these propensities change over time and are coupled with the solution of the ODE system. To circumvent a time-consuming integration of time-dependent propensities and a zero-crossing step as in [7], an additional rejection step is performed. We heuristically choose an integer upper bound for all y_i that is constant for the time until the next discrete reaction (of type $D_1, \ldots$) happens. We use these bounds as the current state of the continuous populations and proceed similar as in Algorithm 1. However, we accept the next discrete reaction (at $t + \tau$) with a probability given by the quotient of the actual propensity at $t + \tau$ and the upper bound of the propensity (computed based on the upper bounds of the y_i's).

In Algorithm 2 we describe the main loop of the simulation procedure. We use the term *current* to refer to those discrete species whose count is non-zero at the current time t of the simulation and explain the details of all subroutines at the end of this section. After the initialization in Line 1, a while-loop is performed that integrates over the time parameter t whose update occurs in Line 8. In Line 3, a flag (which is initially true) is set to false and if at a later step a rejection

1: set $flag = true$ and initialize counts of current discrete and continuous species, counts of all educt patterns and the ODE system;

2: **while** $flag == true$ **do**

3:     set $flag = false$;

4:     for all i, $\overline{y_i}$=determine_upper_bound(y_i) and update the count of all educt patterns accordingly;

5:     **for** each R_i **do**

6:         α_{D_i}=compute_propensity(R_i);

7:     **end for**

8:     compute total propensity $\alpha_D = \sum_i \alpha_{D_i}$;

9:     draw an exponentially distributed pseudo random number τ with rate α_D and update $t = t + \tau$;

10:     simulate the ODE system until $t + \tau$, store the results for all y_i temporarily as y_i^* (restart in Line 4 with higher F if for any i: $y_i^* > \overline{y_i}$);

11:     probabilistically choose a rule R_i according to the propensities α_{D_i};

12:     d_{ij} = instantiate_rule(R_i);

13:     **if** reaction d_{ij} is not possible **then**

14:         set $flag = true$;

15:     **else**

16:         **for** each continuous educt species Y_k of d_{ij} **do**

17:             draw a uniformly $(0,1)$-distributed random number ρ_k;

18:             **if** $\rho_k > y_k^*/\overline{y_k}$ **then**

19:                 set $flag = true$;

20:             **end if**

21:         **end for**

22:     **end if**

23:     set $y_i = y_i^*$ for all continuous species;

24: **end while**

25: **for** all current discrete and continuous species X_i and Y_j **do**

26:     update x_i and y_j according to the chosen reaction;

27: **end for**

Algorithm 2. Main loop of the proposed method: a discrete reaction is rejected when $flag = true$ in Line 13 and Line 19.

happens, the flag is set to true. In Line 4, we heuristically determine an upper bound for all continuous species and update the pattern counts according to that upper bound. In Line 6, the propensities α_{D_i} of the discrete reactions described by rule R_i are computed using the subroutine compute_propensity, which are no longer time-dependent since we use the upper bounds $\overline{y_i}$ instead of the time-varying counts from the solution of the ODE. With these propensities, a time τ until the next attempt of a discrete reaction is computed (Line 9), and the ODE system is integrated until $t + \tau$ using a standard Runge-Kutta-4(5) scheme (Line 10) [12]. Next, a rule R_i is selected with probability α_{D_i}/α_D (Line 11) and for the chosen rule a (combination of) educt species is chosen, using subroutine instantiate_rule (Line 12). In Lines 13–15, we apply the standard rejection step of network-free simulations as in Algorithm 1. Next, we check if the reaction has to be rejected because τ was chosen based on the upper bounds $\overline{y_i}$ and not on the exact (time-varying) propensities (Line 16–21). A uniformly distributed random number ρ is drawn for each continuous educt and compared with the quotient $y_i^*/\overline{y_i}$. If ρ is larger than the quotient, the reaction is rejected and the flag is set to true. We accept a reaction with probability $\prod_{y_i \text{ is educt}} y_i^*/\overline{y_i}$, which is equal to the quotient of the propensities computed based on y_i^* and on $\overline{y_i}$. Note that all propensities follow the law of mass action and rate constants as well as discrete counts cancel in the quotient. If a rejection occurred (flag is true) the loop starts again with the updated values of the y_i's and t.

Subroutine determine_upper_bound. Determining the upper bounds for all continuous species is done heuristically (Line 4). First, the expected time τ^* until the next discrete reaction takes place is computed based on the current total propensities of the discrete reactions (assuming constant propensities). Then by multiplying (only the positive) time-derivatives of the species in Y with τ^* and a factor F, we make a simple overestimation $\overline{y_i}$ for the values of the y_i at time $t + \tau^*$. The value of F is increased if any y_i exceeds $\overline{y_i}$ during the integration of the ODE system. For the three case studies considered in Sect. 4 we found that choosing $F = 30$ does always yield sufficiently high upper bounds while rejections in Line 16–21 are rare.

Subroutine compute_propensity. For computing the propensities of the rules and for instantiating the rules, we track the propensities $\alpha_{c_{ij}}$ for all c_{ij} instead of all $\alpha_{d_{ij}}$, since $|C_i| \ll |D_i|$. This is done by storing explicitly all c_{ij}'s, grouping all c_{ij}'s with equal educts and products, and computing the propensities of these groups. Now, for all rules α_{R_i} is computed as in Algorithm 1 but α_{D_i} is here computed as $\alpha_{R_i} - \sum_j \alpha_{c_{ij}}$. Note that it was not necessary to determine α_{D_i} in Algorithm 1, but here α_{D_i} is needed to instantiate the rules correctly.

Subroutine instantiate_rule. For instantiating a rule, we draw a random number φ that is uniformly distributed between 0 and α_{D_i}. We loop similar as in Algorithm 1, Line 7, over all current species and sum for species $X_1, X_2, \ldots$

$$\sum_{\substack{X_k \text{ matches 1. educt} \\ \text{pattern in } d_{ij}}} \alpha_{d_{ij}} \tag{2}$$

until the sum is greater than φ. Assume that k^* is the smallest index such that the sum is greater than φ. We remark that when summing over all X_k, the result is α_{D_i}. We now select X_{k^*} as the first educt and proceed similar for the second educt if d_{ij} has order two. Note that $\alpha_{d_{ij}}$ only considers those combinations of educt patterns that are *not* already taken into account in the ODE system and therefore, similar as in subroutine compute_propensity, each sum given by Eq. (2) is computed as the difference of the total propensity and the propensities of the according continuous reactions.

Correctness. The proposed hybrid simulation algorithm contains nested rejection steps, because dominating propensities are used instead of the true propensities of the underlying stochastic model. As illustrated in Fig. 2, left, we have the true overall propensity (α, lowest line) and the (possibly higher) propensity $\tilde{\alpha}$ that includes the propensities of reactions described by the rules that are not possible. In addition, we have the dominating upper bound $\overline{\alpha}$ that is given by the upper bounds of the counts of the continuous species. Note that α and $\tilde{\alpha}$ depend on time due to the change of the continuous species.

According to the thinning theorem, the jump times of a non-homogeneous Poisson process with rate $\lambda(t)$ can be sampled by first sampling a Poisson process with rate $\overline{\lambda}(t)$, where $\overline{\lambda}(t) \geq \lambda(t)$, and then successively removing a jump time with probability $1 - \lambda(t)/\overline{\lambda}(t)$. Note that this means that we can iteratively generate exponentially distributed time intervals with rate $\overline{\lambda}(t)$ and successively add them up. To generate a Poisson process with rate $\lambda(t)$ we accept a jump at the end of each interval only with probability $\lambda(t)/\overline{\lambda}(t)$. Both, the standard rejection step of network-free simulations (Line 13/14) and the newly introduced rejection step for the upper bound (Line 16–21) rely on this result. The times τ that are sampled in Line 9 form a Poisson process with rate $\overline{\alpha}$. With probability $1 - \tilde{\alpha}/\overline{\alpha}$ (Line 16–21) we remove jump times such that the remaining times form a process with rate $\tilde{\alpha}$. In addition, we remove jump times with probability $1 - \alpha/\tilde{\alpha}$ (Line 13/14) to finally consider only times forming a process with rate α. Note that in our argumentation, we switched the order of the rejection steps without loss of generality. Note also that the time increment τ and the integration of the ODE is performed in each iteration independent of whether a rejection occurred or not. Finally, we remark that the upper bound $\overline{\alpha}$ may change at the beginning of the loop, which is in accordance with the thinning theorem, because it is also valid for dominating rates that are not constant in time.

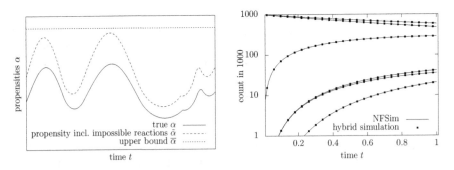

Fig. 2. Left: plot of the propensities that are considered in the proposed algorithm. Right: simulation results of our method compared to standard rule-based simulation (*NFSim*) for the model in Sect. 4.1. The counts are shown for the species A, B, A–B, A–B–A, B–A–B, A–B–A–B, from top to bottom

4   Results

We compare the performance of the proposed hybrid method to a standard rule-based simulation (*NFSim*) for two different polymerization examples. In our implementation we use concentrations instead of average counts for the continuous species[2]. In this way we can easily scale the models by changing the volume and the initial count of the species. Our method is implemented in Go and available at http://mosi.cs.uni-saarland.de/?page_id=1341.

4.1   Theoretical Copolymerization (model 1)

In Fig. 3, we list the details of a simple linear copolymerization that we use to test our method. The reactions follow a step-growth mechanism and the transformation between the monomers was introduced to add a more complex dynamics. Each species that can occur in the system has exactly two sites to perform an elongation reaction. All species are linear chains and the two kinds of monomers will alternate in the species. Note that the given rates are deterministic rates, therefore $k3$ has to be divided by the system size, which we define for simplicity as the count of monomer A at time 0. In Fig. 2, right, we plot the average results of 1000 simulations for a time horizon of 1 sec and an initial count of 10^6 for both A and B. As expected, we get almost identical results for our method and for those computed with *NFSim*. For other models or longer simulation times, the results also perfectly match those of *NFSim*, which means that the mean-field assumption for the continuous species has no significant influence on the average counts when using $K = 200$ as a threshold for continuous species. Therefore, in the sequel we concentrate on a comparison of the running times. At the end of this section, we also show plots for varying K.

$$\text{Monomers: x}-\text{A}-\text{x}, \text{y}-\text{B}-\text{y} \qquad \text{Rules: x}-\text{A}-\text{x} \underset{k2}{\overset{k1}{\rightleftharpoons}} \text{y}-\text{B}-\text{y}$$

$$\text{Rates: } k1 = 0.3, k2 = 0.2$$

$$k3 = 0.2, k4 = 0.13 \qquad \text{A}-\text{x} + \text{y}-\text{B} \underset{k4}{\overset{k3}{\rightleftharpoons}} \text{A}-\text{X}-\text{Y}-\text{B}$$

Fig. 3. Details of the theoretical copolymerization

Since the theoretical copolymerization is intrinsically stiff, i.e. it contains a large number of short chains but only few long chains, we see in Fig. 4 that the running time of the hybrid algorithm (black lines) is significantly shorter than that of *NFSim* (gray lines). Due to the stiffness of the model, the fraction of reactions that can be simulated by solving a system of ODE is high and therefore, the gain in running time is high compared to *NFSim*.

---

[2] Concentrations can be converted to counts and vice versa as described in [28], just by using the factor $V \cdot N_A$, where V is the volume of the system and N_A is Avogadro constant.

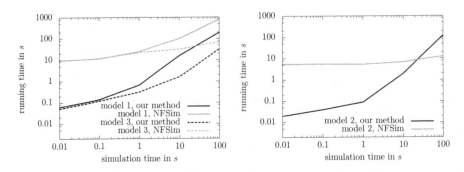

Fig. 4. The running times of the proposed method (black) for different simulation times (time horizons of the model) compared to that of *NFSim* (gray). Left: theoretical copolymerization (model 1, solid lines) and the modified TLBR model (model 3, dashed lines). Right: results of the TLBR model for less stiff parameters (model 2)

4.2 The TLBR Model (Models 2 and 3)

As a second case study, we consider a simple version of the Trivalent Ligand - Bivalent Receptor (TLBR) model proposed in [15, 22] (see Fig. 5). The presented reaction is a copolymerization that uses the step-growth polymerization mechanism. One of the monomers has three reaction sites, which means that during the process polymer networks emerge. In the case of the original parameters, we started the simulations with 500000 ligands, i.e. species B with 3 sites and 30000 receptors. Figure 4, right, shows that the hybrid method is only for short simulation time faster than *NFSim*, because the model is not stiff and almost all reactions that happen in the process are reactions of discrete species. Using modified rates (as given in Fig. 5, left) and using initially 300000 receptor monomers leads to a much stiffer system. Now, the total number of reactions raises by a factor of 10 and most of the reactions are continuous. Therefore, the proposed hybrid simulation method performs better than *NFSim* (Fig. 4, left, dashed lines).

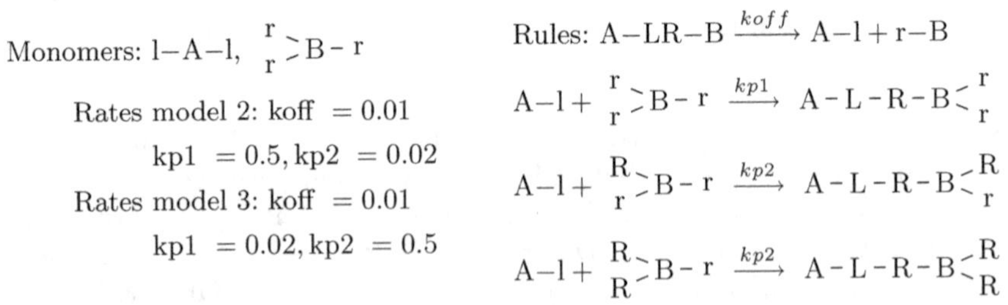

Monomers: $1-A-1$, $\substack{r \\ r} > B - r$ Rules: $A-LR-B \xrightarrow{k_{off}} A-1 + r-B$

Rates model 2: koff $= 0.01$

kp1 $= 0.5$, kp2 $= 0.02$

Rates model 3: koff $= 0.01$

kp1 $= 0.02$, kp2 $= 0.5$

$A-1 + \substack{r \\ r} > B - r \xrightarrow{kp1} A-L-R-B \substack{< r \\ r}$

$A-1 + \substack{R \\ r} > B - r \xrightarrow{kp2} A-L-R-B \substack{< R \\ r}$

$A-1 + \substack{R \\ R} > B - r \xrightarrow{kp2} A-L-R-B \substack{< R \\ R}$

Fig. 5. Reaction scheme and rates of a simple version of the TLBR model. We used two different parameter sets to show principal differences regarding the simulation and the performance

A main feature of the proposed method is that the running time scales well in the system size of the models, which is a major problem in agent-based simulations due to the large amounts of memory needed. In Fig. 6, left, we show results for different system sizes. The running times of the hybrid method is roughly linear in the system size. *NFSim* quits the simulation for a total monomer count of approximately one million due to an out-of-memory error. In Fig. 6, right, we compare the performance of the hybrid simulation for varying K, the threshold for continuous species. In the extreme case that $K = 0$, all counts evolve according to an ODE system. In this case one ODE for each species, also for those with count zero is constructed, i.e. for infinitely many species. We simulated all models (with initial counts for A of 10^6) for a time horizon of one second with different K, starting with $K = 1$. For $K < 20$, it can happen that the method has to reject many times, due to long time intervals until the next discrete reaction happens and therefore less appropriate upper bounds for the y_i's. With $K > 200$ only a few reactions are simulated by ODEs, therefore the stiffness of the models cannot be exploited. We propose to choose K not too small, since even if the means of small populations are approximated well, information about other important measures except the means cannot be estimated anymore. In our case studies, even for small K, the means of small populations were approximated well. However, in general the deterministic mean-field yields only accurate results if the population counts are sufficiently high. We therefore suggest $K \approx 200$ to guarantee accurate results and achieve a considerable speed-up compared to standard methods.

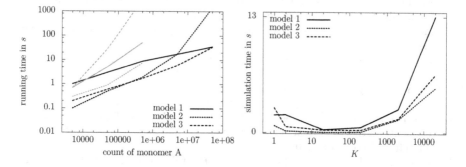

Fig. 6. Left: performance of the proposed method (black) for all three presented models in comparison with *NFSim* (gray). Right: influence of the factor K on the performance of hybrid simulations

5   Conclusion

We proposed a hybrid stochastic simulation algorithm for polymerization reactions. The algorithm is based on a rule-based description of the reactions and performs a network-free simulation, i.e. during the simulation we never explicitly loop over all possible reactions. Moreover, the proposed algorithm treats highly abundant (i.e. continuous) species in a deterministic way by integrating the differential equations of the average changes due to fast (that is, continuous) reactions. The propensities of the remaining discrete reactions depend on the counts of the continuous species and are therefore no longer constant in time. To avoid complicated and inefficient zero-crossing approaches for time-varying propensities, we heuristically choose an upper bound for these propensities until the next slow reaction fires. We exploit the classical thinning approach for Poisson processes and perform a rejection step to compensate for the (on average too small) time step that we chose based on the dominating propensities. For the three case studies that we consider, we see a considerable speedup compared to standard rule-based simulations if the system is stiff and we have almost no losses in accuracy.

Acknowledgments. This work was funded by the Cluster of Excellence on Multimodal Computing and Interaction (MMCI) at Saarland University, Germany.

References

1. Ali Parsa, M., Kozhan, I., Wulkow, M., Hutchinson, R.A.: Modeling of functional group distribution in copolymerization: a comparison of deterministic and stochastic approaches. Macromol. Theory Simul. **23**(3), 207–217 (2014)
2. Anderson, D.F.: A modified next reaction method for simulating chemical systems with time dependent propensities and delays. J. Chem. Phys. **127**(21), 214107 (2007)
3. Bortolussi, L., Krüger, T., Lehr, T., Wolf, V.: Rule-based modelling and simulation of drug-administration policies. In: Proceedings of the Symposium on Modeling and Simulation in Medicine, pp. 53–60. Society for Computer Simulation International (2015)
4. Cao, Y., Gillespie, D.T., Petzold, L.R.: Accelerated stochastic simulation of the stiff enzyme-substrate reaction. J. Chem. Phys. **123**(14), 144917 (2005)
5. Cao, Y., Gillespie, D.T., Petzold, L.R.: The slow-scale stochastic simulation algorithm. J. Chem. Phys. **122**(1), 014116 (2005)
6. Craft, D.L., Wein, L.M., Selkoe, D.J.: A mathematical model of the impact of novel treatments on the aβ burden in the Alzheimers brain, CSF and plasma. Bull. Math. Biol. **64**(5), 1011–1031 (2002)
7. Crudu, A., Debussche, A., Radulescu, O.: Hybrid stochastic simplifications for multiscale gene networks. BMC Syst. Biol. **3**(1), 1 (2009)
8. Danos, V., Feret, J., Fontana, W., Harmer, R., Krivine, J.: Rule-based modelling of cellular signalling. In: Caires, L., Vasconcelos, V.T. (eds.) CONCUR 2007. LNCS, vol. 4703, pp. 17–41. Springer, Heidelberg (2007). doi:10.1007/978-3-540-74407-8_3
9. Danos, V., Feret, J., Fontana, W., Krivine, J.: Scalable simulation of cellular signaling networks. In: Shao, Z. (ed.) APLAS 2007. LNCS, vol. 4807, pp. 139–157. Springer, Heidelberg (2007). doi:10.1007/978-3-540-76637-7_10
10. Danos, V., Laneve, C.: Core formal molecular biology. In: Degano, P. (ed.) ESOP 2003. LNCS, vol. 2618, pp. 302–318. Springer, Heidelberg (2003). doi:10.1007/3-540-36575-3_21
11. Faeder, J.R., Blinov, M.L., Goldstein, B., Hlavacek, W.S.: Rule-based modeling of biochemical networks. Complexity **10**(4), 22–41 (2005)

12. Fehlberg, E.: Low-order classical runge-kutta formulas with stepsize control and their application to some heat transfer problems. Technical report, NASA TR R-315, National Aeronautics and Space Administration, Washington, D.C., July 1969
13. Gillespie, D.T.: Exact stochastic simulation of coupled chemical reactions. J. Phys. Chem. **81**(25), 2340–2361 (1977)
14. Gillespie, D.T.: Stochastic simulation of chemical kinetics. Annu. Rev. Phys. Chem. **58**, 35–55 (2007)
15. Goldstein, B., Perelson, A.S.: Equilibrium theory for the clustering of bivalent cell surface receptors by trivalent ligands. Biophys. J. **45**(6), 1109 (1984)
16. Helal, M., Hingant, E., Pujo-Menjouet, L., Webb, G.F.: Alzheimer's disease: analysis of a mathematical model incorporating the role of prions. J. Math. Biol. **69**(5), 1207–1235 (2014)
17. Herajy, M., Heiner, M.: Hybrid representation and simulation of stiff biochemical networks. Nonlinear Anal. Hybrid Syst. **6**(4), 942–959 (2012)
18. Hogg, J.S., Harris, L.A., Stover, L.J., Nair, N.S., Faeder, J.R.: Exact hybrid particle/population simulation of rule-based models of biochemical systems. PLoS Comput. Biol. **10**(4), e1003544 (2014)
19. Kiparissides, C.: Polymerization reactor modeling: a review of recent developments and future directions. Chem. Eng. Sci. **51**(10), 1637–1659 (1996)
20. Lewis, P.A., Shedler, G.S.: Simulation of nonhomogeneous poisson processes by thinning. Naval Res. Logistics Q. **26**(3), 403–413 (1979)
21. Mastan, E., Zhu, S.: Method of moments: a versatile tool for deterministic modeling of polymerization kinetics. Eur. Polym. J. **68**, 139–160 (2015)
22. Monine, M.I., Posner, R.G., Savage, P.B., Faeder, J.R., Hlavacek, W.S.: Modeling multivalent ligand-receptor interactions with steric constraints on configurations of cell-surface receptor aggregates. Biophys. J. **98**(1), 48–56 (2010)
23. Puchałka, J., Kierzek, A.M.: Bridging the gap between stochastic and deterministic regimes in the kinetic simulations of the biochemical reaction networks. Biophys. J. **86**(3), 1357–1372 (2004)
24. Roland, J., Berro, J., Michelot, A., Blanchoin, L., Martiel, J.L.: Stochastic severing of actin filaments by actin depolymerizing factor/cofilin controls the emergence of a steady dynamical regime. Biophys. J. **94**(6), 2082–2094 (2008)
25. Sneddon, M.W., Faeder, J.R., Emonet, T.: Efficient modeling, simulation and coarse-graining of biological complexity with NFsim. Nat. Methods **8**(2), 177–183 (2011)
26. Thanh, V.H., Priami, C.: Simulation of biochemical reactions with time-dependent rates by the rejection-based algorithm. J. Chem. Phys. **143**(5), 054104 (2015)
27. Van Steenberge, P., Dhooge, D., Reyniers, M.F., Marin, G.: Improved kinetic Monte Carlo simulation of chemical composition-chain length distributions in polymerization processes. Chem. Eng. Sci. **110**, 185–199 (2014)
28. Wolkenhauer, O., Ullah, M., Kolch, W., Cho, K.H.: Modeling and simulation of intracellular dynamics: choosing an appropriate framework. IEEE Trans. Nanobiosci. **3**(3), 200–207 (2004)
29. Wulkow, M.: Numerical treatment of countable systems of ordinary differential equations. Konrad-Zuse-Zentrum für Informationstechnik (1990)
30. Wulkow, M.: Computer aided modeling of polymer reaction engineeringthe status of predici, I-simulation. Macromol. React. Eng. **2**(6), 461–494 (2008)
31. Yang, J., Monine, M.I., Faeder, J.R., Hlavacek, W.S.: Kinetic Monte Carlo method for rule-based modeling of biochemical networks. Phys. Rev. E **78**(3), 031910 (2008)

Model Analysis

Toward Modelling and Analysis of Transient and Sustained Behaviour of Signalling Pathways

Matej Hajnal[1], David Šafránek[1(✉)], Martin Demko[1], Samuel Pastva[1],
Pavel Krejčí[2], and Luboš Brim[1]

[1] Systems Biology Laboratory, Faculty of Informatics, Masaryk University,
Botanická 68a, Brno, Czech Republic
{xhajnal,safranek,xdemko,xpastva,brim}@fi.muni.cz
[2] Department of Biology, Faculty of Medicine, Masaryk University,
Kamenice 735, Brno, Czech Republic
krejcip@med.muni.cz

Abstract. Signalling pathways provide a complex cellular information processing machinery that evaluates particular input stimuli and transfers them into the genome by means of regulation of specific genes expression. In this short paper, we provide a preliminary study targeting minimal models representing the topology of main signalling mechanisms. A special emphasis is given to distinguishing between monotonous (sustained) and non-monotonous (transient) time-course behaviour. A set of minimal parametrised ODE models is formulated and analysed in a workflow based on formal methods.

1 Introduction

Signalling pathways represent one of the most important biochemical mechanisms studied in current systems biology. In particular, they provide a complex cellular information processing machinery that evaluates input stimuli and transfers them into genome by means of regulation of specific genes expression. Due to presence of multiple time scales ranging from nanoseconds to hours, it is difficult to identify the exact dynamics of signalling proteins with a sufficient precision *in vivo* or *in vitro* [18]. Reconstruction of signalling pathways can be thus more challenging than reconstruction of other kinds of biological mechanisms such as metabolic pathways or gene regulatory networks. The increasing interest in signalling pathways is motivated by the fact they are supposed to be the most crucial mechanism in which certain dysfunctions may have cardinal influence on organisms health [15].

A lot of attention has been put to modelling of various kinds of signalling pathways. Besides the traditional ODE models there are applications of Boolean networks [12], process algebraic techniques [8,16,21], hybrid/timed automata or Petri nets [17], all accompanied with relevant formal analysis methods such as

This work has been supported by the Czech Science Foundation grant No. GA15-11089S and by the Czech National Infrastructure grant LM2015055.

E. Cinquemani and A. Donzé (Eds.): HSB 2016, LNBI 9957, pp. 57–66, 2016.
DOI: 10.1007/978-3-319-47151-8_4

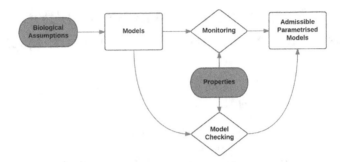

Fig. 1. Formal methods analysis workflow of minimal signalling pathways models

static analysis and model checking. Overview of the methods is provided in [1,5]. In [9], the phenomenon of negative feedback in the common structure of main parts of signalling pathways (MAPK/ERK) is mentioned and experimentally analysed. In [20,22], the authors go even more deeply into details of receptor-activation mechanisms including adaptor proteins and state the hypothesis that negative feedback is chemically implemented via affecting particular parts of receptor-adaptor protein complex conformations. A special emphasis is given to distinguishing between monotonous (*sustained*) and non-monotonous (*transient*) time-course behaviour. It is believed that transition between these two modes may cause a significant change of the nominal cell behaviour leading to serious anomalies of internal cellular processes control.

In this short paper, we provide a preliminary study focused on minimal models representing the topology of signalling mechanisms hypothesised to be present in many organisms [14]. In particular, we formulate a set of minimal parametrised ODE models and analyse them w.r.t. the two distinct time-course behaviour. We employ a workflow based on formal methods. Schematically, the workflow is depicted in Fig. 1. In general, we start with the set of biological assumptions for which we provide a set of ODE models that reflect the level of knowledge on systems kinetics acquired from biophysicists. The models are minimal in the number of variables. Subsequently, we formalise the two types of behaviour in terms of temporal logic properties and employ monitoring and model checking to globally characterise which model variants and their parameterisations satisfy or violate either of the properties. The methodology we employ combines techniques based on simulation and rigorous formal methods. Both kinds of techniques have been successfully employed to extract new information providing deeper understanding of biological mechanisms [2,10].

2   Models of Signalling Pathways

In this section we present individual variants of dynamical models of the considered signalling pathways. The models are based on the following biological assumptions:

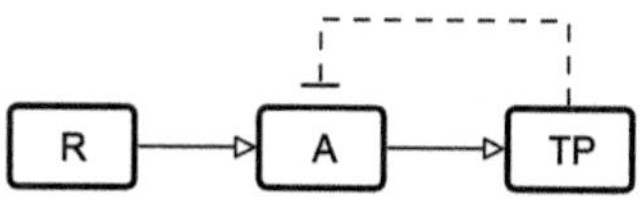

$$hill(X, K_M, n) = \frac{[X]^n}{K_M{}^n + [X]^n} \quad (1)$$

$$hill^-(X, K_M, n) = \frac{K_M{}^n}{K_M{}^n + [X]^n} \quad (2)$$

Fig. 2. Topology of models. Dashed line represents optional inhibition (left). Positive and negative Hill function (right)

1. Main entities of the signalling pathway are signal, receptor, adapter and target protein.
2. There is one signal, receptor, target protein and one major adapter.
3. The input signal is considered saturated in biological experiments, therefore in all models the signal is considered constant.
4. Concentration of the receptor is considered constant. Moreover, with constant signal the activity of the receptor is also constant.
5. The target protein may or may not inhibit the adapter.
6. Inhibition of the adapter may or may not depend on its activation by the receptor.

Based on the assumptions above we construct three models in which we use Hill kinetics as sigmoidal functions (Eq. 1). Since activity of the receptor is constant, there is no dynamical equation required for the input signal. Hence models consist of the following entities: receptor (R), adapter (A) and target protein (TP). Receptor concentration is constant (Eq. 3). The adapter is the main part of the model where non-trivial regulation takes place. It is activated by the receptor and inhibited by the target protein. Model 1 has no inhibition (Eq. 4), Model 2 describes independent inhibition (Eq. 5) and Model 3 describes dependent inhibition (Eq. 6) using negative sigmoidal function (Eq. 2). Target protein dynamics is modelled as a positive sigmoidal function (Eq. 1) of the adapter in all three models (Eq. 7). Resulting topology is illustrated in Fig. 2.

$$\frac{d[R]}{dt} = 0 \quad (3)$$

$$\frac{d[A]}{dt} = V_{MAX\_A} \cdot hill(R, K_{M\_A}, n_A) - y_A[A] \quad (4)$$

$$\frac{d[A]}{dt} = V_{MAX\_A1} \cdot hill(R, K_{M\_A1}, n_{A1}) + V_{MAX\_A2} \cdot hill^-(TP, K_{M\_A2}, n_{A2}) - y_A[A] \quad (5)$$

$$\frac{d[A]}{dt} = V_{MAX\_A1} \cdot hill(R, K_{M\_A1}, n_{A1}) \cdot V_{MAX\_A2} \cdot hill^-(TP, K_{M\_A2}, n_{A2}) - y_A[A] \quad (6)$$

$$\frac{d[TP]}{dt} = V_{MAX\_TP} \cdot hill(A, K_{M\_TP}, n_{TP}) - y_{TP}[TP] \quad (7)$$

With respect to biological assumptions of constant receptor activity, the Hill function of adapter activation is also constant. Therefore, we simplify the model equations. In particular, simplified equations of the adapter for the model with no inhibition, independent inhibition and dependent inhibition, respectively, are the following:

$$\frac{d[A]}{dt} = V_A - y_A[A] \tag{8}$$

$$\frac{d[A]}{dt} = V_A + V_{\text{MAX}\_A} \cdot hill^-(TP, K_{M\_A}, n_A) - y_A[A] \tag{9}$$

$$\frac{d[A]}{dt} = V'_A \cdot hill^-(TP, K_{M\_A}, n_A) - y_A[A] \tag{10}$$

Parameter V'_A is defined as $V_A \cdot V_{\text{MAX\_A}}$. Default parameters have been set to: $V = V'_A = V_{\text{MAX}\_A} = V_{\text{MAX}\_TP} = 0.001$, $K_{M\_A} = K_{M\_TP} = y_A = y_{TP} = 0.1$, $n_A = n_{TP} = 2$. Initial concentrations of all entities have been set to 0: $A(0) = TP(0) = 0$.

3 Analysis Results

The models are analysed w.r.t. the two characteristic types of behaviour formalised in temporal logic. For each model and desired behaviour we try to find parameterisations that satisfy the behaviour or prove its absence. In the most general form, the sustained behaviour can be described by the following template LTLc formula (linear temporal logic with constraints over the reals [11]):

$$\varphi_S = \left(\left(\frac{d[S]}{dt} > 0 \right) \mathbf{U}([S] > a) \right) \wedge (\mathbf{F}(\mathbf{G}([S] > c \cdot a))) \wedge (\mathbf{F}(\mathbf{G}([S] < a))). \tag{11}$$

The transient behaviour can be described by the formula:

$$\varphi_T = \left(\left(\frac{d[S]}{dt} > 0 \right) \mathbf{U}([S] > a) \right) \wedge (\mathbf{F}(\mathbf{G}([S] < c \cdot a))). \tag{12}$$

where S is the observed entity, a is the parameter in the formula template that represents a threshold characterising the concentration bounds on the observed entity behaviour. Additionally, $c \in [0, 1]$ is a constant representing a user-defined scalling factor. The subsequent task is to find model parameter values for which there exists $a > 0$ such that φ_S or φ_T is satisfied.

3.1 Monitoring

First, we employ monitoring analysis in BioCham [19]. Monitoring is applied directly to the ODE models via numerical simulation. To sample the parameter values satisfying a given formula we use the command search_parameters (*list_of-_parameters, list_of_pairs_of_floats, count, φ, time*). The parameter *list_of_para-meters* represents a list of model parameters, *list_of_pairs_of_floats* contains intervals of respective values which are linearly sampled where the number of samples for each parameter is set by the parameter *count*. Model is simulated with sampled parameters in time (0, *time*) until satisfaction of φ is obtained. In order to avoid exponential explosion of samples we have chosen a subset of parameters that have major impact on the stable state of the corresponding differential equation.

Such subset includes production and degradation parameters of each model. To analyse the sustained behaviour, we employ LTL formula φ'_S that represents a stronger condition than φ_S. Instead of φ_T we employ weaker LTL formula φ'_T describing necessity of transient behaviour. The relations among the formulas are illustrated in Fig. 3 (right). The modified formulae are the following:

$$\varphi'_S = \mathbf{G}(\frac{d[TP]}{dt} >= 0), \tag{13}$$

$$\varphi'_T = \mathbf{F}(\frac{d[TP]}{dt} < -0.0001). \tag{14}$$

Following results are summarised in Fig. 3 (left).

Model 1

Sustained behaviour has been analysed with the following command:

$$\texttt{search\_parameters}([V_A, y_A, V_{\text{MAX\_TP}}, y_{TP}], [(0.001, 100)^*], 10, \varphi'_S, 1000). \tag{15}$$

Monitoring has shown that the initial parametrisation is a witness for the sustained behaviour.
Transient behaviour has been analysed with the following command:

$$\texttt{search\_parameters}([V_A, y_A, V_{\text{MAX\_TP}}, y_{TP}], [(0.001, 100)^*], 10, \varphi'_T, 1000). \tag{16}$$

Monitoring has resulted with "No value found". This formula describes necessity of transient behaviour, therefore, we are not able to find transient behaviour by monitoring. To verify or refute presence of transient behaviour we will use model checking (Sect. 3.2).

Model 2

Sustained behaviour has been analysed with the following command:

$$\texttt{search\_parameters}([V_A, V_{\text{MAX\_A}}, y_A, V_{\text{MAX\_TP}}, y_{TP}], [(0.001, 100)^*], 10, \varphi'_S, 1000). \tag{17}$$

Monitoring has shown that the initial parametrisation is a witness for the sustained behaviour.
Transient behaviour has been analysed with the following command:

$$\texttt{search\_parameters}([V_A, V_{\text{MAX\_A}}, y_A, V_{\text{MAX\_TP}}, y_{TP}], [(0.001, 100)^*], 10, \varphi'_T, 1000). \tag{18}$$

The decrease of TP has not been sufficient, thus, we have used additional parameter $PEEK$ with interval [0,1] within the following formula:

$$\varphi''_T = \mathbf{F}([TP] > PEEK) \wedge \mathbf{F}(\mathbf{G}([TP] < PEEK \cdot 0.55)), \tag{19}$$

which results in parameterisation:

$(V_A, 0.001), (V_{\text{MAX\_A}}, 0.5009), (y_A, 0.5009), (V_{\text{MAX\_TP}}, 2.5005), (y_{TP}, 0.5009), (PEEK, 1).$

	sustained	transient
Model 1	yes	no
Model 2	yes	yes
Model 3	yes	yes

$$\varphi_S''$$
|
$$\varphi_S'$$
|
$$\varphi_S$$

$$\varphi_T$$
|
$$\varphi_T'$$
|
$$\varphi_T''$$

Fig. 3. Table of monitoring results (left). Lattices of used LTL formulae with meaning *True* if a satisfying parameterisation has been found or *False* if not (right). The upper formula represents a stricter condition

Model 3

Sustained behaviour has been analysed with the following command:

$$\texttt{search\_parameters}([V_A', y_A, V_{\text{MAX\_}TP}, y_{TP}], [(0.001, 100)^*], 10, \varphi_S', 1000). \quad (20)$$

Simulation of resulting parametrisation has shown low concentration of TP in comparison with concentration of A. To distinguish resulting behaviour from constant zero, we have used the following formula:

$$\varphi_S'' = \mathbf{G}(\frac{d[TP]}{dt} >= (0)) \wedge \mathbf{F}([TP] > d), \quad (21)$$

where d has been set proportionally to concentration of A, $d = 0.1$. That way we have obtained more realistic parameterisation:

$$(V_A', 0.001), (y_A, 0.001), (V_{\text{MAX\_}TP}, 0.05095), (y_{TP}, 0.05095).$$

Transient behaviour has been analysed with the following command:

$$\texttt{search\_parameters}([V_A', y_A, V_{\text{MAX\_}TP}, y_{TP}], [(0.001, 100)^*], 10, \varphi_T', 1000). \quad (22)$$

Monitoring has shown that the initial parametrisation is a witness for the transient behaviour:

$$(V_A', 0.05095), (y_A, 0.001), (V_{\text{MAX\_}TP}, 0.05095), (y_{TP}, 0.001).$$

3.2   Model Checking Analysis

In order to prepare the model for model checking analysis, we have first constructed the piece-wise affine approximation (PAA) of the original non-linear continuous models. To this end, we have subsequently applied the automatised approximation and abstraction procedures introduced in [3, 13]. In particular, we have approximated each non-linear function (i.e., Hill function) appearing in the right-hand side of the model equations with a sum of ten piece-wise affine ramp functions optimally fitting the original kinetic function. As a result of abstraction, we have obtained a parameterised direction transition system (PDTS)

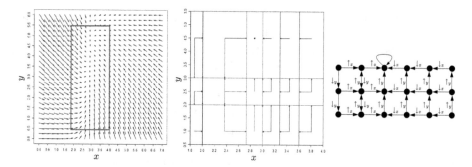

Fig. 4. A vector field of an approximated ODE system (left) and discretisation of the emphasised region (middle). Thresholds determining the rectangles were obtained by the algorithm in [13]. Arrows in the rectangles show the directions of transitions abstracting the systems dynamics in the particular rectangle. States and transitions of the labelled transition system corresponding to the discretisation (right)

that exactly over-approximates the PAA. Details on abstracting the models into respective PDTSs can be found in [7]. In Fig. 4, there is shown the main principle of the abstraction. Additionally, the transitions are naturally labelled by an up- (resp. down-) arrow expressing the change in particular model variable.

Properties are reformulated in terms of UCTL formulae [4]. UCTL is a branching-time temporal logic that combines action-based and state-based approach. The logic is able to express state predicates over system states, action predicates over transitions, and combine these with temporal and Boolean operators in the style of CTL. We employ event predicates to express requirements on the character of species concentration change, the so-called directions, as shown in Fig. 4 (right). The usage of branching time logic is motivated by the fact that the rectangular abstraction produces a non-deterministic system. Combination of action and state predicates is necessary to express local patterns of the dynamics (state predicates) and the character of the transitions (action predicates).

We employ the parameter synthesis based on coloured CTL model checking extended to deal with UCTL operators. Since we restrict ourselves to particular operators as needed in the properties of our interest, the extension is a direct refinement of the algorithm described in [6]. The procedure gives us a global result saying for which initial states and respective parameter values the formula is satisfied. It is important to note that the over-approximating abstraction affects interpretation of the parameter synthesis results. On one hand, satisfaction is guaranteed for universally quantified formulae. On the other hand, falsification is guaranteed for existentially quantified formulae. Quantifiers cannot be alternated. In both cases of guaranteed results, the obtained parameter values are under-approximated.

Parameters for synthesis are chosen as follows: V_A in Model 1, $V_{\text{MAX\_A}}$ in Model 2, V'_A in Model 3 and $V_{\text{MAX\_TP}}$ in all three models. These parameters are examined in the range [0.0001,10] and other constants are used according to Sect. 2 except y_A, y_{TP} (set to 0.5009) and V_A in Model 2 (set to 0.001).

To express combined characteristics of the two behaviour of TP, we have employed the following properties in the form of UCTL formulae:

$$\varphi_1 = Init \ \wedge \ _\uparrow\mathbf{AX}(_{\neg\downarrow}\mathbf{AF} \ Stable),$$
$$\varphi_2 = Init \ \wedge \ \mathbf{EF}(_\uparrow\mathbf{EX}(_{\neg\downarrow}\mathbf{EF}(_{\neg\uparrow}\mathbf{EF}(_\downarrow\mathbf{EX} \ True)))),$$
$$\varphi_3 = Init \ \wedge \ \mathbf{EF}(_\uparrow\mathbf{EX}(_{\neg\downarrow}\mathbf{EF}(_{\neg\uparrow}\mathbf{EF}(_\downarrow\mathbf{EX} \ NotUp)))),$$

where $Init$ stands for the set of initial states (with TP constrained in the range [0.0,0.0001]), $NotUp$ for a state in which the abstracted vectorfield forbids an increase in TP concentration, $Stable$ for an equilibrium state with both species being stable and, finally, $True$ for any state. Intuitively, property φ_1 specifies the sustained behaviour and its satisfaction guarantees presence of this behaviour in a given model for given initial states. Formally, this formula restricts all admissible runs to start in $Init$ and to increase TP concentration in the very next step and not to decrease it before reaching $Stable$. Properties φ_2 and φ_3 represent necessary conditions for presence of the transient behaviour (formulae are ordered with respect to the strength of the particular condition). Both restrict any feasible run to start in $Init$ and to increase TP concentration at least once on a run before decreasing it also at least once on a run. Parameterisations violating these properties guarantee absence of the transient behaviour. All results are summarised in Table 1.

It is important to note that the abstraction together with UCTL logic allow us to exactly characterise inevitability of the sustained behaviour. However, as it has been reported in [7], the state predicate $Stable$ is exactly preserved in the abstraction only if a particular equilibrium that must exist in the respective state rectangle is hyperbolic. In consequence, we are not able to fully cover the tran-

Table 1. The results obtained for given properties in different models. Initial concentration of A is defined as a union of concentration values in states where the formula holds for stated parameters. V^* is the production parameter of A that represents V_A in Model 1, V_{MAX_A} in Model 2, V_A' in Model 3, respectively. Each parameter interval determines a range of all satisfying parameter values across all states of the respective model where the particular property holds

Model type	Property	Initial concentration of A	$V^* \times V_{MAX\_TP}$
Model 1	φ_1	[0.22,11.9]	[0.11,5.96]×[0.0,6.81]
	$\neg\varphi_2$	[0.22,12.0]	[0.11,10.0]×[7.23,10.0]
	$\neg\varphi_3$	[0.01,12.0]	[0.23,10.0]×[6.26,10.0] ∪[5.96,10.0]×[0.0,10.0]
Model 2	φ_1	[0.01,11.5]	[0.0,5.76]×[0.0,0.01]
	$\neg\varphi_2$	[0.22,12.0]	[2.29,10.0]×[7.23,10.0]
	$\neg\varphi_3$	[0.0,12.0]	[0.09,10.0]×[7.23,10.0] ∪[4.47,10.0]×[0.0,10.0]
Model 3	φ_1	[0.01,11.5]	[0.01,5.76]×[0.0,0.01]
	$\neg\varphi_2$	[0.22,12.0]	[2.31,10.0]×[7.23,10.0]
	$\neg\varphi_3$	[0.0,12.0]	[0.11,10.0]×[7.23,10.0] ∪ [4.49,10.0]×[0.0,10.0]

sient behaviour. More specifically, the transient behaviour might asymptotically converge to an equilibrium that is asymptotically stable. Such a property is not preserved by the abstraction and therefore we limit ourselves only to refuting *absolutely transient* behaviour (transient behaviour without oscillations around the target equilibrium).

4   Conclusions

We have provided a preliminary study toward analysis of dynamics of two types of behaviour characteristic for signalling pathways. We started by directly analysing the set of minimal models by monitoring. This has given us partial results on which model variants provide either of the behaviour. In particular, for each of the three models we have found a parameterisation that produces the sustained type of behaviour. These results have been further confirmed by model checking to be guaranteed for certain ranges of parameter values. In the case of the transient type of behaviour, positive results have been found by monitoring only for two models. For the model without the negative feedback, no positive result has been found provided that we cannot conclude the answer.

Further analysis by model checking has given us parameterisations for which the absence of the transient behaviour is guaranteed. Although these results are affected by the approximation error, they give a good preliminary knowledge for future design of targeted biological experiments focused on concrete types of signalling pathways. Deeper exploration of how the two techniques – underapproximative monitoring and overapproximative model checking – can be combined to obtain a general automatised analysis framework is the goal of our future work.

References

1. Bartocci, E., Liò, P.: Computational modeling, formal analysis, and tools for systems biology. PLoS Comput. Biol. **12**(1), 1–22 (2016)
2. Bartocci, E., Liò, P., Merelli, E., Paoletti, N.: Multiple verification in complex biological systems: the bone remodelling case study. In: Priami, C., Petre, I., Vink, E. (eds.) Transactions on Computational Systems Biology XIV. LNCS, vol. 7625, pp. 53–76. Springer, Heidelberg (2012). doi:10.1007/978-3-642-35524-0_3
3. Batt, G., Belta, C., Weiss, R.: Model checking liveness properties of genetic regulatory networks. In: Grumberg, O., Huth, M. (eds.) TACAS 2007. LNCS, vol. 4424, pp. 323–338. Springer, Heidelberg (2007)
4. ter Beek, M.H., Fantechi, A., Gnesi, S., Mazzanti, F.: A state/event-based model-checking approach for the analysis of abstract system properties. Sci. Comput. Program. **76**, 119–135 (2011)
5. Brim, L., Češka, M., Šafránek, D.: Model checking of biological systems. In: Bernardo, M., de Vink, E., Di Pierro, A., Wiklicky, H. (eds.) SFM 2013. LNCS, vol. 7938, pp. 63–112. Springer, Heidelberg (2013)
6. Brim, L., Češka, M., Demko, M., Pastva, S., Šafránek, D.: Parameter synthesis by parallel coloured CTL model checking. In: Roux, O., Bourdon, J. (eds.) CMSB 2015. LNCS, vol. 9308, pp. 251–263. Springer, Heidelberg (2015)

7. Brim, L., Demko, M., Pastva, S., Šafránek, D.: High-performance discrete bifurcation analysis for piecewise-affine dynamical systems. In: Priami, C., Petre, I., de Vink, E. (eds.) HSB 2015. LNCS, vol. 9271, pp. 58–74. Springer, Heidelberg (2015). doi:10.1007/978-3-319-26916-0_4

8. Calder, M., Duguid, A., Gilmore, S., Hillston, J.: Stronger computational modelling of signalling pathways using both continuous and discrete-state methods. In: Priami, C. (ed.) CMSB 2006. LNCS (LNBI), vol. 4210, pp. 63–77. Springer, Heidelberg (2006)

9. Courtois-Cox, S., Williams, S.M.G., Reczek, E.E., Johnson, B.W., McGillicuddy, L.T., Johannessen, C.M., Hollstein, P.E., MacCollin, M., Cichowski, K.: A negative feedback signaling network underlies oncogene-induced senescence. Cancer Cell 10(6), 459–472 (2006)

10. Donzé, A., Clermont, G., Langmead, C.J.: Parameter synthesis in nonlinear dynamical systems: application to systems biology. J. Comput. Biol. 17(3), 325–336 (2010)

11. Fages, F., Rizk, A.: On the analysis of numerical data time series in temporal logic. In: Calder, M., Gilmore, S. (eds.) CMSB 2007. LNCS (LNBI), vol. 4695, pp. 48–63. Springer, Heidelberg (2007)

12. Grieco, L., Calzone, L., Bernard-Pierrot, I., Radvanyi, F., Kahn-Perls, B., Thieffry, D.: Integrative modelling of the influence of mapk network on cancer cell fate decision. PLoS Comput. Biol. 9(10), 1–15 (2013)

13. Grosu, R., Batt, G., Fenton, F.H., Glimm, J., Le Guernic, C., Smolka, S.A., Bartocci, E.: From cardiac cells to genetic regulatory networks. In: Gopalakrishnan, G., Qadeer, S. (eds.) CAV 2011. LNCS, vol. 6806, pp. 396–411. Springer, Heidelberg (2011)

14. Kholodenko, B.N.: Cell-signalling dynamics in time and space. Nat. Rev. Mol. Cell Biol. 7(3), 165–176 (2006)

15. Klipp, E., Liebermeister, W.: Mathematical modeling of intracellular signaling pathways. BMC Neurosci. 7(1), 1–16 (2006)

16. Kwiatkowska, M.Z., Heath, J.K.: Biological pathways as communicating computer systems. J. Cell Sci. 122(16), 2793–2800 (2009)

17. Li, C., Suzuki, S., Ge, Q.W., Nakata, M., Matsuno, H., Miyano, S.: Structural modeling and analysis of signaling pathways based on Petri nets. J. Bioinform. Comput. Biol. 04(05), 1119–1140 (2006)

18. Li, X., Shen, L., Shang, X., Liu, W.: Subpathway analysis based on signaling-pathway impact analysis of signaling pathway. PLoS ONE 10(7), 1–19 (2015)

19. Rizk, A., Batt, G., Fages, F., Soliman, S.: A general computational method for robustness analysis with applications to synthetic gene networks. Bioinformatics 25(12) (2009)

20. Sasagawa, S., Ozaki, Y.I., Fujita, K., Kuroda, S.: Prediction and validation of the distinct dynamics of transient and sustained ERK activation. Nat. Cell Biol. 7(4), 365–373 (2005)

21. Wang, D.Y., Cardelli, L., Phillips, A., Piterman, N., Fisher, J.: Computational modeling of the EGFR network elucidates control mechanisms regulating signal dynamics. BMC Syst. Biol. 3(1), 1–17 (2009)

22. Yamada, S., Taketomi, T., Yoshimura, A.: Model analysis of difference between EGF pathway and FGF pathway. Biochem. Biophys. Res. Commun. 314(4), 1113–1120 (2004)

Application of the Reachability Analysis for the Iron Homeostasis Study

Alexandre Rocca[1,2]([✉]), Thao Dang[1], Eric Fanchon[2], and Jean-Marc Moulis[3]

[1] VERIMAG/CNRS, 700 Avenue Centrale, 38400 Saint Martin D'Hères, France
alexandre.rocca@imag.fr
[2] Université Grenoble-Alpes - Grenoble 1/CNRS, TIMC-IMAG,
UMR 5525, 38041 Grenoble, France
[3] Université Grenoble-Alpes - Grenoble 1, Laboratoire de Bioénergétique
Fondamentale et Appliquée (LBFA) - Inserm U1055, Grenoble, France

Abstract. Our work is motivated by a model of the mammalian cellular Iron Homeostasis, which was analysed using simulations in [9]. The result of this analysis is a characterization of the parameters space such that the model satisfies a set of constraints, proposed by biologists or coming from experimental results. We now propose an approach to hypothesis validation which can be seen as a complement to the approach based on simulation. It uses reachability analysis (that is set-based simulation) to formally validate a hypothesis. For polynomials systems, reachability analysis using the Bernstein expansion is an appropriate technique. Moreover, the Bernstein technique allows us to tackle uncertain parameters at a small cost. In this work, we extend the reachability analysis method presented in [7] to handle polynomial fractions. Furthermore, to tackle the complexity of the Iron Homeostasis model, we use a piecewise approximation of the dynamics and propose a reachability method to deal with the resulting hybrid dynamics. These approximations and adaptations allowed us to validate a hypothesis stated in [9], with an exhaustive analysis over uncertain parameters and initial conditions.

Keywords: Parametric ODE · Reachability analysis · Non-linear systems · Biological systems

1 Introduction

In cellular biology, models are often based on the elementary law of chemical reactions, or empirical laws for lumped reactions, and expressed in terms of Ordinary Differential Equations (ODEs). However, unlike the models of classical chemistry, most of the parameters in the biological models are uncertain, or can greatly vary from one sample (or one individual) to another. For these reasons, modelling in biology is often a round trip between experimental results and validation of a hypothesis about a biological mechanism formulated by a model. These models being uncertain, hypothesis validation is often done with numerous numerical simulations. However, this simulation-based method is both costly

© Springer International Publishing AG 2016
E. Cinquemani and A. Donzé (Eds.): HSB 2016, LNBI 9957, pp. 67–84, 2016.
DOI: 10.1007/978-3-319-47151-8_5

for large parameter spaces, and not exhaustive. Formal verification techniques allow proving properties, with set-based reachability computation techniques, by replacing simulation runs with conservative sets of trajectories. In this paper, we study a model of the mammalian cellular Iron Homeostasis, which was analysed using simulations in [9]. The result of this analysis is a characterization of the parameters space such that the model satisfies a set of constraints, proposed by biologists or coming from experimental results. We now propose an approach which can be seen as a complement to the approach based on simulations. It uses reachability analysis (that is set-based simulation) to formally validate a hypothesis on the model. For polynomials systems, reachability analysis using the Bernstein expansion is an appropriate technique. Moreover, the Bernstein technique allows us to tackle uncertain parameters at a small cost. In this work, we extend the reachability analysis method presented in [7] to handle polynomial fractions. Furthermore, to tackle the complexity of the Iron Homeostasis model, we use a piecewise approximation of the dynamics and propose a reachability method for the resulting hybrid dynamics. These approximations and adaptations allowed us to validate a hypothesis stated in [9], with an exhaustive analysis over uncertain parameters and initial conditions. The paper is organized as follows. We first formulate the problem and describe the model of Iron Homeostasis. We then recall the Bernstein technique [7] and present our adaptations and approximations. Finally, we report the experimental results.

2   The Mammalian Cellular Iron Homeostasis (MCIH) Model

The model we consider is based on the model of Iron Homeostasis proposed in [9]. This is an ODE model built to study and represent the mechanism of Iron Homeostasis for a large parameter space. The work [9] provides a method to characterize a large valid parameter space (19 parameters, spanning several orders of magnitude), by finding the parameters points which respect a set of temporal constraints and clustering them in multiple ellipsoids. Here we define our experiments based on both the model and the results from this work.

This model describes the control of the iron concentration inside a cell, thanks to both an iron storage protein, ferritin, and regulating proteins the two IRP. Moreover, both the transferrin receptor TfR1 (which influences the iron input in the cell) and the iron exporting protein FPN1a are influenced by the IRP concentration. Tf_{sat} is the external saturated transferrin concentration, which is the iron transport protein outside the cell. The concentration of free iron in the cell that is not stored inside ferritin must be well controlled since too much or too little of it can have deleterious effects.

In the presence of a stable concentration of iron-loaded transferrin, outside the cell, the cell state converges to a steady state. When there is no more iron outside the cell (Tf_{sat} is almost equal to 0) the non-ferritin bound iron quickly drops for a short time, but then increases again at the expense of ferritin iron and stabilizes for some time (around 10 h) as the regulation mechanism activates.

The low iron concentration stimulates the IRP activity which itself activates the release of the iron stored in the ferritin. This supply of iron from the ferritin leads to a pseudo-steady state for a few hours, until the ferritin concentration is too low to release enough iron to maintain the equilibrium. If no iron is added to the medium shortly thereafter, the cell dies.

The equations of the ODE model representing this mechanism are the following. The model contains 5 state variables (Fe, IRP, Ft, TfR1, FPN1a), and 19 parameters, and the equations are the following.

$$\frac{dFt}{dt} = kFt_{prod} - k_{IRP-Ft} \cdot sig(IRP, \theta_{IRP-Ft}, d_{sig}) - kFt_{deg} \cdot Ft$$

$$\frac{dFe}{dt} = kFe_{input} \cdot Tf_{sat} \cdot TfR1 - nFt \cdot \frac{dFt}{dt} - kFe_{export} \cdot Fe \cdot FPN1a - kFe_{cons} \cdot Fe$$

$$\frac{dIRP}{dt} = kIRP_{prod} - k_{Fe-IRP} \cdot sig(Fe, \theta_{Fe-IRP}, d_{sig}) \cdot IRP - kIRP_{deg} \cdot IRP$$

$$\frac{dFPN1a}{dt} = kFPN1a_{prod} - k_{IRP-FPN1a} \cdot sig(IRP, \theta_{IRP-FPN1a}, d_{sig}) - kFPN1a_{deg} \cdot FPN1a$$

$$\frac{dTfR1}{dt} = kTfR1_{prod} + k_{IRP-TfR1} \cdot IRP - kTfR1_{deg} \cdot TfR1$$

where $sig(x, \theta, d_{sig}) = \dfrac{x^{d_{sig}}}{x^{d_{sig}} + \theta^{d_{sig}}}$.

The work [9] made the observation that for all the valid parameter points, the value of FPN1a is almost not influenced by the value of the other variables during the experiments, FPN1a being almost constant with this modelling. FPN1a, being the iron exporting protein, should quickly decrease with the iron concentration Fe, as modelled by the IRP dependency. However, the FPN1a concentration stays stable for all the *valid* parameters points.

In this work, we propose, as a proof of concept of our method, to compute the reachable set of FPN1a for the parameters [kFPN1a$_{deg}$, kFPN1a$_{prod}$, k$_{IRP-FPN1a}$, $\theta_{IRP-FPN1a}$] taken in the interval given by the valid parameter points. The computed reachable set must ensure that in presence of external iron, the system evolves to a steady state, and in absence of external iron, there is a plateau of at least $10\,h$ for the variable Fe, followed by a decrease in iron concentration.

3   Hypothesis Validation Problem Using Reachability Analysis

Our objective is to validate a hypothesis on some properties of a given biological model, which contains uncertain parameters. Furthermore, the initial conditions, such as the initial concentrations of most species in the MCIH, are not accurately known but they lie within identified intervals. Note that a particularity of biological models is that the sets of parameter values (as well as the set of initial conditions in some cases) can be large. Therefore, the hypothesis should be validated for all the behaviours generated by such nondeterminism, for which reachability analysis is an appropriate tool.

This method was developed for polynomial systems and is based on the Bernstein expansion of polynomials [7] whereas the dynamics of our model contains terms involving polynomial fractions, which are used to capture switching phenomena. A direct extension of the method for polynomial fractions on one hand may cause significant errors which may forbid deriving meaningful results. On the other hand the procedure can be computationally expensive because of the high degrees of the polynomials required to accurately model the switching dynamics. To address these issues, we propose combining two ideas: hybridization and an approximation by fractions of polynomials with lower degrees; the former is used to retain the switching behaviours and the latter to reduce computational complexity. Another adaptation concerns an application of implicit form of the Bernstein expansion in order to address the sparsity of the polynomials often encountered in biological models.

In the following we first recall the main ideas of a reachability analysis method and we then describe the adaptations.

3.1   Reachable Set Computation Using the Bernstein Expansion

We consider a discrete-time polynomial dynamical system of the form:

$$x[k+1] = \pi(x[k], p),$$
$$x[0] \in X_0$$

where $x \in \mathbb{R}^n$ is the state variables, $\pi(x, p)$ is a parametric n-variate polynomial, p is a parameter vector taking values in the parameter space $\mathcal{P}$, X_0 is the initial set. The interest of such models in the analysis of biological models lies in the fact that many biological models are constructed using data measured in discrete time. To handle the model of Iron Homeostasis, which is a continuous-time model, we will use a discrete-time approximation of this model obtained using the Euler discretization, which for this particular case is sufficiently accurate to retain important behaviours observed at discrete time points. Note however that the question of discretization error is crucial and should be considered for general cases.

We now introduce the notion of reachable set. Given a set $X \subset \mathbb{R}^n$ of states, the image of X by π for all parameter values $p \in \mathcal{P}$ is

$$\pi(X, \mathcal{P}) = \{\pi(x, p) \mid x \in X \wedge p \in \mathcal{P}\}.$$

The reachable set from a set X after k steps, denoted by **Reach**$_k(X)$ is the set X_k computed by the following iterations:

$$X_k = \pi(X_{k-1}, \mathcal{P}),$$
$$X_0 = X$$

In this work, we use the reachability analysis method developed in [7]. Because the exact image $\pi(X_k, \mathcal{P})$ is not easy to compute, an over-approximation of the image $\pi(X_k, \mathcal{P})$ is computed as a polyhedron defined by a set of m linear constraints $\Gamma x \leq d$. In particular, the matrix Γ of size $n \times m$ is fixed a priori and such polyhedra are called template polyhedra, denoted by $\langle \Gamma, d \rangle$. To compute this over-approximation we need to compute the offset vector d such that $\forall x \in \pi(X_k, \mathcal{P})$ we have $\Gamma x \leq d$. Let Γ_i be the ith line of the constraint matrix Γ, we search for d_i such that $\forall x \in \pi(X_k, \mathcal{P})$ we have $\Gamma_i x \leq d_i$, which can be written as the following optimization problem:

$$d_i = max\{\Gamma_i \pi(x, p) \mid x \in X_k \wedge p \in \mathcal{P}\}. \tag{1}$$

With the offset vector d determined as above, the polyhedron $\langle \Gamma, d \rangle$ is guaranteed to contain the image $\pi(X_k, \mathcal{P})$.

The Bernstein expansion can be used to compute enclosures of a polynomial and is an efficient method to provide an approximate solution to the above maximization problem that needs to be solved at each step of the reachability computation. In the following we briefly explain the Bernstein expansion of a polynomial. We use multi-indices of the form $\mathbf{i} = (\mathbf{i}_1, \mathbf{i}_2, \dots, \mathbf{i}_n)$ where each $\mathbf{i}_j$ is a non-negative integer. Given two multi-indices $\mathbf{i}$ and $\mathbf{l}$, we write $\mathbf{i} \leq \mathbf{l}$ if for all $j \in \{1, \dots, n\}$, $\mathbf{i}_j \leq \mathbf{l}_j$. Also, we write $\binom{\mathbf{i}}{\mathbf{l}}$ for $\binom{\mathbf{i}_1}{\mathbf{l}_1} \binom{\mathbf{i}_2}{\mathbf{l}_2} \dots \binom{\mathbf{i}_n}{\mathbf{l}_n}$. Let $\mathcal{B}$ denote the unit box anchored at the origin, that is $\mathcal{B} = [0, 1]^n$. A polynomial of degree $\mathbf{l} = (\mathbf{l}_1, \dots, \mathbf{l}_n)$ is often written using a monomial representation:

$$\pi(x) = \sum_{\mathbf{i} \leq \mathbf{l}} a_{\mathbf{i}}(p) x^{\mathbf{i}}$$

where $x^{\mathbf{i}} = x_1^{\mathbf{i}_1} x_2^{\mathbf{i}_2} \dots x_n^{\mathbf{i}_n}$ and the parameter p appears in the coefficients of the monomials. For all $x \in \mathcal{B}$, $\pi(x)$ can also be represented using the following Bernstein expansion:

$$\pi(x) = \sum_{\mathbf{i} \leq \mathbf{l}} b_{\mathbf{i}}(p) B_{\mathbf{l}, \mathbf{i}}(x)$$

where $b_{\mathbf{i}}(p)$ are the Bernstein coefficients (of degree $\mathbf{l}$) of π, and $B_{\mathbf{l}, \mathbf{i}}(x)$ are the Bernstein basis polynomials: $B_{\mathbf{l}, \mathbf{i}}(x) = \beta_{\mathbf{l}_1, \mathbf{i}_1}(x_1) \dots \beta_{\mathbf{l}_n, \mathbf{i}_n}(x_n)$ where for a real number y we have $\beta_{\mathbf{l}_j, \mathbf{i}_j}(y) = \binom{\mathbf{l}_j}{\mathbf{i}_j} y^{\mathbf{i}_j} (1 - y)^{\mathbf{l}_j - \mathbf{i}_j}; j \in \{1, \dots, n\}$. The Bernstein coefficients are given by the following formulas which can be easily computed:

$$b_{\mathbf{i}}(p) = \sum_{\mathbf{j} \leq \mathbf{i}} \frac{\binom{\mathbf{i}}{\mathbf{j}}}{\binom{\mathbf{l}}{\mathbf{j}}} a_{\mathbf{j}}(p), 0 \leq \mathbf{i} \leq \mathbf{l}.$$

The enclosure property of the Bernstein expansion we are interested in is the following: for all $p \in \mathcal{P}$ and for all $x \in \mathcal{B}$

$$\pi(x, p) \subseteq \square(\{b_{\mathbf{i}}(p) \mid \mathbf{i} \leq \mathbf{l}\})$$

where $\square$ is the operator that computes the bounding box of a set of points.

We have briefly explained the Bernstein expansion for the polynomial π. In the remainder of the paper we manipulate a number of different polynomials, we thus add the name of the polynomial as a superscript in the notation of the Bernstein coefficients, for example b_j^π. Note that the above Bernstein expansion is valid only for the unit box. Therefore, if X_k is the unit box, let γ_i be the linear function over x: $\gamma_i(x) = \Gamma_i x$ and $\eta = \pi \circ \gamma_i$, that is $\eta(x, p) = \gamma_i(\pi(x, p))$. Now we represent the polynomial $\eta(x, p)$ by its Bernstein expansion and let $b_{max}^\eta(p) = \max_{i \leq l}(b_i^\eta(p))$, the optimization problem (1) becomes

$$d_i = max\{b_{max}^\eta(p) \mid p \in \mathcal{P}\}. \tag{2}$$

If $\mathcal{P}$ is a convex polyhedron and p appears linearly in the monomial coefficients of π, the above is a linear program (LP). The offset vector d can thus be determined by solving m linear programs and the polyhedron described by $\Gamma x \leq d$ is an over-approximation of the image $\pi(X_k, \mathcal{P})$.

However, even if the initial set is the unit box, after the first iteration, the set X_k becomes a polyhedron (with the template matrix Γ). To extend the above idea to polyhedral sets, we first remark that the idea can be easily extended for box-like domains, such as arbitrary non-axis-aligned boxes, or parallelotopes, by composing the polynomial of the dynamics with an affine function describing the transformation from the unit box to the box-like domain in question. In order to generalize this approach further to handle polyhedral sets in the optimization problem (1), in [7] a method for decomposing a convex polyhedron X_k into a set of parallelotopes $\{\Pi_j \mid j \in \{1, \ldots, m_\Pi\}\}$, called a parallelotopic bundle, was proposed. More concretely, a parallelotopic bundle represents a polyhedron which is the intersection of all the parallelotopes in the bundle; hence we can write $X_k = \Pi_1 \cap \ldots \cap \Pi_{m_\Pi}$. Thus, to over-approximate the image of X_k, it suffices to compute the image of each parallelotope in its bundle and then take their intersection, that is

$$\pi(X_k, \mathcal{P}) \subseteq \pi(\Pi_1, \mathcal{P}) \cap \ldots \cap \pi(\Pi_{m_\Pi}, \mathcal{P}). \tag{3}$$

3.2   Adaptations for the Analysis of the MCIH Model

Because of the complexity of the MCIH model, we need to improve the method proposed by [7] and use approximations of the model of [9]. The first improvement concerns the computation of the Bernstein coefficients. The number of Bernstein control points is of order $\mathcal{O}((|1| + 1)^n)$ and is thus exponential in the degree of the polynomial and the number of variables. Expanding over the maximal degree is therefore not efficient for sparse polynomials which contains only few monomial terms involving few variables. This is often the case in biological models (where the evolution of the concentration of a substance depends only on the concentrations of few other substances). An implicit representation of the Bernstein coefficients, proposed in [13], considers each monomial separately. The expansion of each monomial has thus less Bernstein coefficients to compute.

In addition, the expansion of a monomial can be computed efficiently. For example, given a monomial $\mu = \mu_1 \mu_2 \ldots \mu_n$ where each $\mu_j = x_j^{i_j}$ $(0 \leq j \leq n)$ is a univariate monomial, the Bernstein coefficients of μ (of degree $\mathbf{l} = (l_1, \ldots, l_n)$) are the product: $b_{\mathbf{i}} = \prod_{j=1}^{n} b_{i_j}^{\mu_j}$ where $b_{i_j}^j$ is the $\mathbf{i}_j^{th}$ Bernstein coefficient (of degree l_j) of the univariate monomial $m_j = x_j^{i_j}$.

With the model under study we need to optimize not only polynomials but also fraction of polynomials. In [10] a method to obtain an enclosure of rational functions over box domains is proposed. If π and π' are two polynomials from $\mathbb{R}^n$ to $\mathbb{R}$, and $q = \dfrac{\pi}{\pi'}$, then [10] states that the enclosure of $q(x)$ for all x inside the unit box is given by:

$$min_{\mathbf{i}}(b_{\mathbf{i}}^{\pi}/b_{\mathbf{i}}^{\pi'}) \leq q(x) \leq max_{\mathbf{i}}(b_{\mathbf{i}}^{\pi}/b_{\mathbf{i}}^{\pi'})$$

Note that we can simply compute an upper bound by taking the maximum of the numerator and the minimum of the denominator, but this bound is however much coarser approximations. Note also that we need to compute all the Bernstein coefficients if we want to keep a good approximation of the enclosure.

The original model of Iron Homeostasis leads to a huge number of Bernstein coefficients because of the high degree of the sigmoids which are rational functions. On the other hand, simply lowering the degree d_{sig} can cause significant errors, compared to the original model. To cope with this difficulty, each sigmoid has been approximated by a piecewise function. For a sigmoid function $\dfrac{x^{d_{sig}}}{x^{d_{sig}} + \theta^{d_{sig}}}$ (where x and θ are scalar variables), the associated piecewise function of x and θ (d_{sig} being a constant) is:

$$sig(x, \theta, d_{sig}) = \begin{cases} 0, & \text{if } x \leq \dfrac{(d_{sig} - 2)\theta}{d_{sig}} \\ \dfrac{d_{sig}(x - \theta)}{4\theta} + \dfrac{1}{2}, & \text{if } x > \dfrac{(d_{sig} - 2)\theta}{d_{sig}} \text{ and } x \leq \dfrac{(d_{sig} + 2)\theta}{d_{sig}} \\ 1, & \text{if } x > \dfrac{(d_{sig} + 2)\theta}{d_{sig}} \end{cases}$$

The MCIH model contains 4 parameters taken on large intervals. While the parameters kFPN1a$_{deg}$, kFPN1a$_{prod}$, and k$_{IRP-FPN1a}$ appear linearly in the dynamics, $\theta_{IRP-FPN1a}$ appears non-linearly. For this reason, we treat the parameter $\theta_{IRP-FPN1a}$ as the sixth variable, and thus the term $\dfrac{IRP^5}{IRP^5 + \theta_{IRP-FPN1a}^5}$ is not approximated by a piecewise linear function but its approximation is non-linear in $\theta_{IRP-FPN1a}$ in one piece. These piecewise approximations lead to a new model where each sigmoid is substituted by a 3-piece approximation. In place of one ODE system, the dynamics is now hybrid with 15 modes (locations)[1]. To deal with this hybrid dynamics, we need an efficient way to encode this partition of the state space, so that it is easy to locate a set on the partition during its evolution under the dynamics. To this end, we use Binary Space Partition Tree (BSPT)

---

[1] Two sigmoids are on IRP, and one on Fe.

techniques [8]. Each node of a BSPT is associated with a domain defined by a set of linear constraints, each constraint h is of the form $\alpha \cdot x \le \beta$ where $\alpha \in \mathbb{R}^n$ and $\beta \in \mathbb{R}$. We denote by $\neg h$ the constraint $\alpha \cdot x > \beta$. For simplicity of notation and presentation we use h to denote both the constraint and the half-space defined by this constraint, and instead of saying that the domain of a node intersects with some set, we simply that say a node intersects with some set.

Let $H = \{h_i \mid i \in \{1, \ldots, m\}\}$ be the set of all constraints in the above piece-wise approximations of the sigmoid functions in the dynamics. The domain associated with the root node $\top$ of the tree is the whole state space. For each leaf node, if adding one constraint from H splits the corresponding domain into two sub-domains, we create two child nodes from it, each corresponds to a sub-domain. This constraint is called the *splitting constraint* of the node. We repeat the same procedure until all the constraints in H are added. An example of the construction of the BSPT for the approximation of $sig(\text{IRP}, \theta_{\text{IRP-Ft}})$, $sig(\text{Fe}, \theta_{\text{Fe-IRP}})$ (labelled with $sig_{\text{IRP-Ft}}$ and $sig_{\text{Fe-IRP}}$ in the figure) can be seen in Fig. 1. We associate with the root node the highest rank m, and a child node has a rank smaller than its parent by 1. Once the BSPT is constructed, each leaf of this tree corresponds to a location the domain of which is a convex polyhedron.

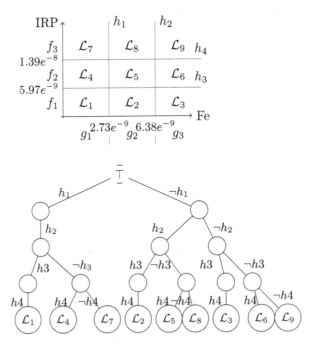

Fig. 1. First picture: 9 locations generated by $[f_1; f_2; f_3] = [0; 1.25e^8 \cdot \text{IRP} - \frac{3}{4}; 1]$ the approximation of $sig(\text{IRP}, \theta_{\text{IRP-Ft}})$ and $[g_1; g_2; g_3] = [0; 2.74e^8 \cdot \text{Fe} - \frac{3}{4}; 1]$ the one of $sig(\text{Fe}, \theta_{\text{Fe-IRP}})$. The constraints h_i are of the form $\alpha \cdot x \le \beta$, for example h_1: Fe $\le 2.73e^{-9}$. Second picture: the BSPT representing the partition, $\top$ is the whole state space.

We now explain how to locate a given polyhedral set X on the partition, that is identify the locations the domains of which intersect with X. Note that such sets X are generated by the dynamics of the system, we can exploit its continuity to perform this operation efficiently. Since the approximate system under study is continuous, we can expect that the successor set X_{k+1} is in the same locations or in the locations which are adjacent to the current locations. Thus, during the reachability process, instead of starting the search from the root of the tree, we can start from the adjacent locations. We call these starting nodes the guess nodes. At the initial step $k = 0$, since there is no information, we start our search from the root of the tree. The search algorithm, named locating, consists of the following two steps.

In the first step, we search for the node $\mathcal{N}$ of lowest rank which strictly contains X, that is $X \subseteq \mathcal{N}$. To do so, we test if there is no node in the guess list satisfying this condition. If this is the case, we go upward in the tree and test their parent nodes, until the condition is satisfied. If the node $\mathcal{N}$ found this way is a leaf, $\mathcal{N}$ is the only location containing X, the algorithm returns $\mathcal{N}$ and stops. Otherwise, it proceeds to the second step starting from $\mathcal{N}$. As an example, if X_k is in the location $\mathcal{L}_5$, then X_{k+1} intersects both $\mathcal{L}_5$ and $\mathcal{L}_8$ (Fig. 1)[2]: if the guess list contains only $\mathcal{L}_5$, the lowest rank node which contains X_{k+1} is the parent of $\mathcal{L}_5$. One can easily see that the efficiency of this algorithm depends on the ordering of the constraints when building the BSPT. In any case using guess lists is not worse than starting from the root.

The second step of the algorithm is breadth-first search starting from the node $\mathcal{N}$, in order to obtain $(\mathcal{L}, \mathcal{X})$ where $\mathcal{L}$ is the set of locations with non-empty intersections, and $\mathcal{X}$ is the set of corresponding intersections (stored as a set of convex polyhedra). This step is a recursive procedure applied to a set $\mathcal{L}$ of nodes intersecting with X, until we reach the leaves. Initially, $\mathcal{L} = \{\mathcal{N}\}$, and the set $\mathcal{X}$ contains only $\{X\}$. Then, until $\mathcal{L}$ contains only leaves when the algorithm returns $(\mathcal{L}, \mathcal{X})$ and then stops, the following procedure is iterated. As we shall see, by construction, at any iteration, all the nodes in $\mathcal{L}$ have the same rank and therefore have the same splitting constraint, denoted by h:

- If $X \subseteq h$, then X intersects the left child of every node in $\mathcal{L}$, then $\mathcal{L} = $ left_children($\mathcal{L}$).
- If $X \subseteq \neg h$, then X intersects the right child of every node in $\mathcal{L}$, then $\mathcal{L} = $ right_children($\mathcal{L}$).
- If X satisfies none of the two above conditions, then X intersects both all the right and left children of the nodes in $\mathcal{L}$. Then, $\mathcal{L} = $ left_children($\mathcal{L}$) $\cup$ right_children($\mathcal{L}$), and the algorithm updates $\mathcal{X} = \{\mathcal{X} \cap h\} \cup \{\mathcal{X} \cap \neg h\}$, where $\{\mathcal{X} \cap h\}$ and $\{\mathcal{X} \cap \neg h\}$ are polyhedra resulting from intersecting each set in $\mathcal{X}$ with the half-spaces corresponding to the constraints h and $\neg h$ respectively.

For each polyhedral intersection in $\mathcal{X}$, it is now possible to compute an associated bundle using the decomposition technique in [7]. Algorithm 1 summarizes a single step of reachability computation, given a fixed template Γ, a BSPT Ψ and its

---

[2] This situation corresponds to the phase 3 described in Sect. 4.

Algorithm 1. Reach(X,Ψ,Γ,$\mathcal{P}$)

1: INPUT: X: Current set (template polyhedron)
2: INPUT: Ψ: BSPT and its associated piecewise approximation of the dynamics.
3: INPUT: Γ: template which is used to over-approximate the reachable set.
4: INPUT: $\mathcal{P}$: parameters set
5: OUTPUT: an over-approximation of the reachable set of X after one step.

6: $\mathcal{G}$: set of current locations, or the tree root $\top$

7: $(\mathcal{L},\mathcal{X}) = \Psi.\texttt{locating}(X,\mathcal{G})$ /* finding intersecting locations */

8: **for** $\Gamma_i \in \Gamma$ **do**
9:     /* computing offset bound d_i for each constraint Γ_i in the template */

10:    **for** $(\mathcal{L}_\kappa, X_\kappa) \in \mathcal{L}$ **do**
11:        Compute the approximate dynamics f_κ associated with the location $\mathcal{L}_\kappa$
12:        **for** $\Pi_j \in \texttt{Bundle}(X_\kappa)$ **do**
13:            /* decompose polyhedron X_κ into a parallelotopic bundle */

14:            Construct the polynomial $\eta_{i,\kappa,j}$ from polynomial f_κ of the dynamics,
15:            template constraint Γ_i and paralletopic domain Π_j, as defined in Sect. 3

16:            $d_{i,\kappa,j} = \max\{\eta_{i,\kappa,j}(x,p)) \mid x \in \mathcal{B} \wedge p \in \mathcal{P}\}$
17:            /* using the Bernstein expansion for polynomial $\eta_{i,\kappa,j}$ */

18:        **end for**

19:        $d_{i,\kappa} = \min_j(d_{i,\kappa,j})$
20:        /* smallest bound over all parallelotopes Π_j in the bundle */

21:    **end for**

22:    $d_i = \max_\kappa(d_{i,\kappa})$
23:    /* largest bound by all approximate dynamics of intersecting locations */

24: **end for**
25: **return** $\langle \Gamma, d \rangle$ /* the result is the template polyhedron with offsets d */

---

associated piecewise approximations. To compute the reachable set at time t from an initial set X_0, with a fixed time step δ, we iterate the Algorithm 1 $\frac{t}{\delta}$ times. In this algorithm, we compute an offset bound for each template constraint Γ_i as follow. We first find all the intersection of the current set with the location domains. Then, for each intersecting location $\mathcal{L}$, we decompose the intersection into a parallotopic bundle, and then for each parallotope, we find a bound on the composed polynomial (with respect to the template constraint Γ_i and to the parallelotope). We do the same for the other parallelotopes and take the minimal bound (corresponding to the intersection of the images shown in (3)). Now we do the same for all the approximate dynamics of the intersecting locations and then take the largest bound, in order to guarantee a conservative approximation of the reachable set.

4   Experiments and Results

4.1   Experimentation Methodology

Recall that our goal is to validate the observations, which were obtained in [9] using numerical simulations, about the regulation of FPN1a. These observations had been done with parameter values chosen such that the system respects some properties:

- In presence of external iron input ($Tf_{sat} \neq 0$), the Fe, Ft, and IRP concentrations reached a steady state.
- In absence of external iron input ($Tf_{sat} = 0$), the iron concentration first stabilized on a plateau for at least 10h, then decreased to 0.

The reachability analysis produces an over-approximation of the reachable set. Because of accumulated error, this set may grow at each step in every directions. We thus do not impose strong constraints for the 'plateau' definition, and currently restrict to a qualitative observation. For the same reason and because we are interested in the question whether the FPN1a concentration strongly decreases during the $Tf_{sat} = 0$ phase, we restrict to a qualitative observation on the lower bound of the reachable set of FPN1a. The reachability analysis of the adapted model was done using the following method: starting from initial conditions (taken from [9]) and a corresponding valid parameter p, these initial conditions are bloated to a set. The following parameters $[kFPN1a_{deg}, kFPN1a_{prod}, k_{IRP-FPN1a}, \theta_{IRP-FPN1a}]$ are extended to cover a few orders of magnitude based on the results of [9]. From this starting initial set, we first let the system evolve to a steady state with $Tf_{sat} \neq 0$. This is the mode where the system should be stable. We let the system stabilize for a few hours. Some results of this stabilization are shown on Figs. 2 and 3. It is clear that the system evolves towards an invariant set, and converges. Because this tool does not compute a precise invariant set, we will take, for the next part of the computation, this over-approximation as the initial set. The initial set for this part is given in Table 1 in Appendix.

In the second part of the analysis we reduce Tf_{sat} to 0: this is the mode where the external iron is depleted. We simulate then 32 h (230400 iterations using a fixed time step of 0.5 s) of the iron depleted mode.

4.2   Observations

In Figs. 4, 5 and 6, we can observe the different phases of the computations in different colors (blue, red, green, and purple).

− Phase 1 (blue): On the initial state previously computed, we apply the following change: Tf_{sat} drop from 1 to 0. Experimentally this corresponds to washing the external medium of the cell and replacing it by a medium without iron. This sudden change of Tf_{sat} leads to the very low iron concentration at t near 0 (see Fig. 4). This very low iron concentration activates the production of IRP, which

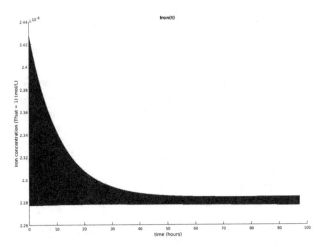

Fig. 2. Iron stabilization

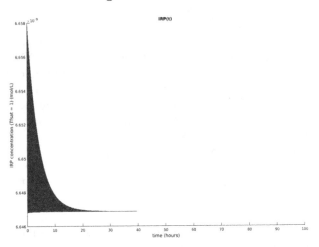

Fig. 3. IRP stabilization

itself activates the release of iron by the ferritin. The iron and the IRP concentrations quickly grow back until both the IRP and iron are around their respective thresholds θ_{IRP-Ft} and θ_{Fe-IRP}. The IRP increase slows down while the iron concentration stabilizes. To compute precisely this blue part, reachability analysis was done using 15 different directions to represent template polyhedral set. The reachability computation time for the blue part is around 2 h.

– Phase 2 (red): Because the iron is now in the wanted plateau, we need to ensure that the analysis is as precise as possible. Thus, we reduce the error by bisecting the set on the IRP axis, and perform reachability analysis with two smaller sets instead of one big set. Even with such a method, one can observe

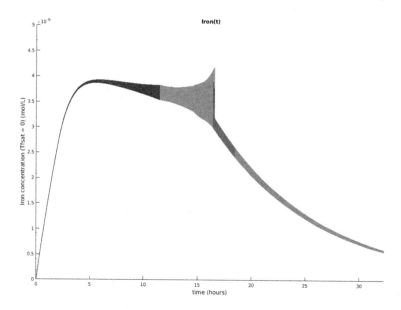

Fig. 4. Iron reachable set for $\text{Tf}_{\text{sat}} = 0$ (Color figure online)

the fast growing accumulated error in the red phase. In the red part, the system overlaps two partitions of IRP: the one where $sig(\text{IRP}, \theta_{\text{IRP}-\text{Ft}})$ is represented by an affine function, and the one where $sig(\text{IRP}, \theta_{\text{IRP}-\text{Ft}}) = 1$. Overlapping two partitions increases the error during a short time, leading to the observed growth of the reachable set in red in Fig. 4.

– Phase 3 (green): Once the reachable set has completely crossed the border between the two partitions, and $sig(\text{IRP}, \theta_{\text{IRP}-\text{Ft}}) = 1$, the reachable set quickly contracts, and the iron concentration begins to decrease notably. Reciprocally, the IRP concentration increases trying to compensate the lack of iron. However at this moment, there is no longer enough ferritin to supply the cell in iron. The red part and the green part took around 3 h to compute in total.

– Phase 4 (purple): The iron concentration is not stable in a plateau, but now decreases to 0. To compute this part we did not need as good precision as before and used a simple box over-approximation, and the computation time of the purple part is around 15 min.

4.3 Discussion

The reachability analysis of the system allows us to validate the previous observation made in [9] using point-based simulations: the regulation term of FPN1a by only IRP in this model and within these parameters intervals is not effective. This suggests that another actor is needed for the regulation of FPN1a. Indeed, even with the large initial set for FPN1a, and having the parameters influencing

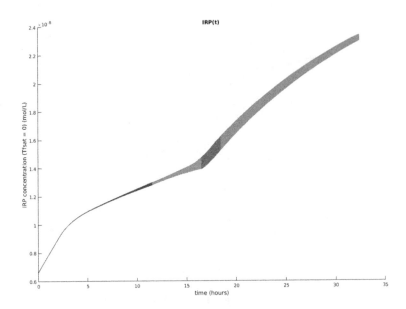

Fig. 5. IRP reachable set for $\text{Tf}_{\text{sat}} = 0$ (Color figure online)

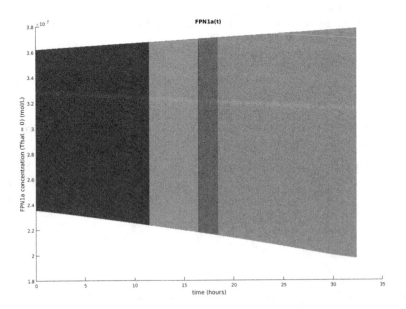

Fig. 6. FPN1a reachable set for $\text{Tf}_{\text{sat}} = 0$ (Color figure online)

FPN1a spanning over large intervals (multiple orders of magnitude), the reachability analysis results show that the model satisfies the expected properties:

1. Fe, IRP, and Ft tend to a small invariant set when $Tf_{sat} \neq 0$.
2. The iron concentration reaches a plateau for at least $10\,h$.
3. After reaching a plateau the iron concentration decreases to 0.
4. The IRP concentration first increases quickly then more slowly during the plateau and then increases quickly again.

However given all those conditions the FPN1a concentration did not undergo any notable decrease. Indeed, in Fig. 6, while the upper bound slowly increase due to the accumulated error, the lower bound, which is conservative, does not decrease notably unlike what we could expect.

This analysis shows that if the model efficiently represents the regulation of the iron concentration with the IRP proteins, it does not fully model the FPN1a regulation, and and IRP is not the main regulating factor in this regime on the FPN1a concentration.

5   Related Works

To study biochemical systems such as the MCIH model presented in this work, we based our framework on the C++ tool Sapo [7] for polynomial systems. The work [7] showed that Bernstein is an efficient way to compute reachable sets of polynomial parametric ODE models. As polynomials in biological models are often sparse, using the implicit formulation of the Bernstein expansion in [13] allows avoiding the explosion of Bernstein coefficients with the dimension and thus improving reachability computation for biological models. Moreover, the Sapo tool allows us to handle uncertain parameters, and could be used in the future for parameter synthesis. Adding the improved computation of the Bernstein coefficients and the approximation by piece-wise non-linear models, the current framework can perform reachability analysis on a large class of biological models with switching behaviours. Our work can be compared to other works which focus on reachability analysis of non-linear biological models.

There are similar previous works using the Bernstein expansion, such as [6,12]. The work [6] allows performing reachability analysis over polytopic sets, instead of bundles of parallelotopes. However, this approach does not directly handle parametric models and is much slower than [7] due to the conversion from polytopes to boxes. The work [12] uses the Bernstein expansion to compute an LP-relaxation of a polynomial optimization problem (POP), which is then solved over a polytopic set. If this technique can compute reachable sets with high precision with polytopes, it is more expensive than [7] which only needs to compute the parametric Bernstein coefficients once.

Some other tools such as [1–3] have been developed mainly for reachability analysis of biological models. The work in [2,3] is dedicated to piecewise multi-affine models with either conical representations of reachable sets or rectangular abstractions. While [1] focuses on the parameters synthesis of piecewise multi-affine models such as gene networks, our MCIH model is more complex. The piecewise approximation is our work is similar to the hybridization in [5].

Finally we can mention well-known tools such as `Flow*` [4] and `Keymaera` [11], for the reachability analysis of non-linear systems. `Flow*` is an efficient tool based on Taylor models for approximating flowpipes in form of boxes, while `Sapo` computes flowpipes constituted of polytopes. `Flow*` can be used for more general non-linear hybrid models but it does not seem to extend easily to parametric analysis. The tool `Keymaera` [11] uses a different approach: it is a theorem prover based on differential logic. It requires knowing solutions to differential equations or solving them numerically. It can compute invariants; however in the context of systems biology, it is very useful when interacting with biologists to provide temporal flowpipes.

6   Conclusion

In this work reachability analysis has been used to validate some hypotheses previously proposed in [9]. The studied model of the Iron Homeostasis is a parametric polynomial ODE: to perform reachability analysis on this model we proposed adaptations of the method based on the Bernstein expansion and piecewise approximation of the dynamics. The resulting method is thus well adapted to biological models having polynomial dynamics with uncertain parameters and switching behaviors. We validated the hypothesis developed in [9] on the modelling of the FPN1a regulation, and this analysis can be seen as the proof of concept of our method. The reachability analysis using the Bernstein decomposition is a fast, and yet precise enough method to perform reachability analysis on parametric polynomial biological system. Further improvements include automatic procedures for set splitting and choosing the template. Also, as future work we will focus on taking advantage of the dynamical properties of the system, such as multiple time-scales or local stability. These properties could allow to speed up computations by transposing the problem in a locally equivalent problem of lower dimension. They could also improve the precision, giving hints for the template directions or the splitting choice.

Acknowledgement. This work is partially supported by the ANR CADMIDIA project (ANR-13-CESA-0008-03) and the ANR MALTHY project (ANR-12-INSE-003).

Appendix

Table 1. Initial Set after Stabilization on the left and parameter Space on the right.

parameters	Value or Interval	Unit
$kFPN1a_{deg}$	$[1e^{-7} \quad 1e^{-6}]$	s^{-1}
$kFPN1a_{prod}$	$[1e^{-17} \quad 1e^{-13}]$	$mol \cdot (L \cdot s)^{-1}$
$k_{IRP-FPN1a}$	$[1e^{-17} \quad 1e^{-13}]$	$mol \cdot (L \cdot s)^{-1}$
$\theta_{IRP-FPN1a}$	$[1e^{-8} \quad 2.01e^{-6}]$	$mol \cdot L^{-1}$
k_{Fe-IRP}	$5.24e^{-5}$	s^{-1}
kFe_{cons}	$1.56e^{-1}$	s^{-1}
kFe_{export}	$2.191e^{3}$	$L \cdot (mol \cdot s)^{-1}$
kFe_{input}	$3.65e^{-2}$	s^{-1}
kFt_{deg}	$2.92e^{-5}$	s^{-1}
kFt_{prod}	$8.93e^{-12}$	$mol \cdot (L \cdot s)^{-1}$
k_{IRP-Ft}	$8.71e^{-12}$	$mol \cdot (L \cdot s)^{-1}$
$k_{IRP-TfR1}$	$3.03e^{-4}$	s^{-1}
$kIRP_{deg}$	$1.5e^{-5}$	s^{-1}
$kIRP_{prod}$	$4.48e^{-13}$	$mol \cdot (L \cdot s)^{-1}$
$kTfR1_{deg}$	$2.23e^{-5}$	s^{-1}
$kTfR1_{prod}$	$1.78e^{-13}$	$mol \cdot (L \cdot s)^{-1}$
θ_{Fe-IRP}	$9.89e^{-9}$	$mol \cdot L^{-1}$
θ_{IRP-Ft}	$4.56e^{-9}$	$mol \cdot L^{-1}$
nFt	177.4	$-$
d_{sig}	5	$-$

Variable	Interval	Unit
Fe	$[2.27e^{-8} \quad 2.28e^{-8}]$	mol/L
IRP	$[6.646e^{-9} \quad 6.647e^{-9}]$	mol/L
Ft	$[2.804e^{-7} \quad 2.805e^{-7}]$	mol/L
FPN1a	$[2.35e^{-7} \quad 3.61e^{-7}]$	mol/L
TfR1	$[9.8e^{-8} \quad 10.2e^{-8}]$	mol/L

References

1. Batt, G., Yordanov, B., Weiss, R., Belta, C.: Robustness analysis and tuning of synthetic gene networks. Bioinformatics **23**(18), 2415–2422 (2007)
2. Berman, S., Halász, Á., Kumar, V.: MARCO: a reachability algorithm for multi-affine systems with applications to biological systems. In: Bemporad, A., Bicchi, A., Buttazzo, G. (eds.) HSCC 2007. LNCS, vol. 4416, pp. 76–89. Springer, Heidelberg (2007). doi:10.1007/978-3-540-71493-4_9
3. Brim, L., Fabriková, J., Drazan, S., Safranek, D.: Reachability in biochemical dynamical systems by quantitative discrete approximation (2011). arXiv preprint: arXiv:1107.5924
4. Chen, X., Ábrahám, E., Sankaranarayanan, S.: Flow*: an analyzer for non-linear hybrid systems. In: Sharygina, N., Veith, H. (eds.) CAV 2013. LNCS, vol. 8044, pp. 258–263. Springer, Heidelberg (2013)
5. Dang, T., Maler, O., Testylier, R.: Accurate hybridization of nonlinear systems. In: Proceedings of the 13th ACM International Conference on Hybrid Systems: Computation and Control, pp. 11–20. ACM (2010)
6. Dang, T., Testylier, R.: Reachability analysis for polynomial dynamical systems using the bernstein expansion. Reliable Comput. **17**(2), 128–152 (2012)

7. Dreossi, T., Dang, T., Piazza, C.: Parallelotope bundles for polynomial reachability. In: Proceedings of the 19th International Conference on Hybrid Systems: Computation and Control, pp. 297–306. ACM (2016)

8. Fuchs, H., Kedem, Z.M., Naylor, B.F.: On visible surface generation by a priori tree structures. ACM Siggraph Comput. Graph. **14**, 124–133 (1980). ACM

9. Mobilia, N.: Méthodologie semi-formelle pour l'étude de systèmes biologiques: application à l'homéostasie du fer. Ph.D. thesis, Université Joseph Fourier, Grenoble (2015)

10. Narkawicz, A., Garloff, J., Smith, A.P., Munoz, C.A.: Bounding the range of a rational functiom over a box. Reliable Comput. **17**, 34–39 (2012)

11. Platzer, A., Quesel, J.-D.: KeYmaera: a hybrid theorem prover for hybrid systems (system description). In: Armando, A., Baumgartner, P., Dowek, G. (eds.) IJCAR 2008. LNCS (LNAI), vol. 5195, pp. 171–178. Springer, Heidelberg (2008)

12. Sassi, M.A.B.: Analyse et contrôle des systèmes dynamiques polynomiaux. Ph.D. thesis, Université de Grenoble (2013)

13. Smith, A.P.: Fast construction of constant bound functions for sparse polynomials. J. Glob. Optim. **43**(2–3), 445–458 (2009)

Synchronous Balanced Analysis

Andreea Beica[✉] and Vincent Danos

École Normale Supérieure, Rue d'Ulm 45, 75005 Paris, France
beica@di.ens.fr

Abstract. When modeling Chemical Reaction Networks, a commonly used mathematical formalism is that of Petri Nets, with the usual interleaving execution semantics. We aim to substitute to a Chemical Reaction Network, especially a "growth" one (i.e., for which an exponential stationary phase exists), a piecewise synchronous approximation of the dynamics: a resource-allocation-centered Petri Net with maximal-step execution semantics. In the case of unimolecular chemical reactions, we prove the correctness of our method and show that it can be used either as an approximation of the dynamics, or as a method of constraining the reaction rate constants (an alternative to flux balance analysis, using an emergent formally defined notion of "growth rate" as the objective function), or a technique of refuting models.

Keywords: Chemical Reaction Networks · Approximation · Resource allocation · Max-parallel execution of Petri Nets · Flux balance analysis

1 Introduction

When studying certain cellular processes, the assumption is that the remainder of the cell can either be ignored or considered constant. Despite this assumption, intracellular processes rarely work in isolation, but rather in continuous interaction with the rest of the cell. Furthermore, the cell has finite resources, so committing resources to one task reduces the amount of resources available to others. All cells experience these trade-offs, which potentially modify all cellular processes, but are often overlooked. In this paper, we propose a piecewise synchronous approximation of the dynamics of Chemical Reaction Networks (CRN) based on finite resource allocation between reactions, that puts these trade-offs front stage. One goal is to rephrase the mass action run of the system as a problem of optimization: the inter-phase between synchronous runs defines an unknown, the resource split, and we can ask for the best split (e.g., the one which minimizes parallel completion time, or maximizes growth rate). Our method allows us to define a formal notion of growth rate for our type of Petri Net execution, that can serve as an improved "biomass objective function" [11] for a constraint method similar to flux balance analysis (FBA) [11]: "Synchronous Balanced Analysis".

Related work. While most intracellular growth processes are well characterized, the manner in which they are coordinated under the control of a scheduling

© Springer International Publishing AG 2016
E. Cinquemani and A. Donzé (Eds.): HSB 2016, LNBI 9957, pp. 85–94, 2016.
DOI: 10.1007/978-3-319-47151-8_6

policy is not well understood. When fast replication is sought, a schedule that minimizes the completion time is naturally desirable. But when resources are scarce, in the worst case it is computationally hard to find such a schedule [1,2]. The scheduling problem of a self-replicating bacterial cell is studied in [3]. A mathematical cell model that respects the resource trade-offs experienced by cells is built in [4]. The concept of maximally parallel execution already appears in the literature on P-systems [5], and in Levy's family reductions [6], while in [7], the authors use it to develop a Petri Net execution semantics that resembles biology. The scheduling policy of cells is also tackled in [8], where the notion of *bounded asynchrony* is introduced. In [9], the author introduces a constraint method that generalizes FBA to the stochastic case, allowing models to be discriminated using second order moments.

This paper is organised as follows. The next section contains an overview of how CRNs can be modeled using Petri Nets (PNs), as well as our definition of max-parallel execution of a PN. Next we introduce our piecewise synchronous execution semantics and show it encompasses max-parallel execution. We then demonstrate that, at least in the case of unimolecular reactions, it recreates the usual ODE system dynamics, and that it can be used either as an approximation of the dynamics, or as an alternative to flux balance analysis. The final section concludes with a summary and outlook regarding further work on the subject.

2 Modeling Chemical Reaction Networks

Definition 1. *A chemical reaction network (CRN) is a tuple $< \mathcal{S}, \nabla^-, \nabla^+, \mathcal{R}, \kappa >$, where $\mathcal{S} = \{S_1, \ldots, S_s\}$ is a finite set of species, ∇^- and ∇^+ are $r \times s$ consumption, respectively production stoichiometry matrices, $\mathcal{R} = \{r_1, \ldots, r_r\}$ is a finite set of reactions, and $\kappa : \mathcal{R} \to \mathbb{R}_{>0}$ associates a (positive) rate constant to each reaction.*

Each reaction $r_i \in \mathcal{R}$ is of the form $\sum_{j=1}^s \nabla_{ij}^- S_j \xrightarrow{k_i} \sum_{j=1}^s \nabla_{ij}^+ S_j$, and the reaction network can be written compactly in matrix-vector form as $\nabla^- S \xrightarrow{k} \nabla^+ S$, with S the species column vector, and k the rates column vector.

The state of a system can be represented as a multiset of the concentrations of all the chemical species in the network, denoted by $\mathbf{x} = (\mathbf{x}_{S_1}, \ldots, \mathbf{x}_{S_s}) \in \mathbb{N}^s$. Applying the law of mass action, the dynamics of the reaction network assumed to be in state $\mathbf{x}$ are given by the kinetic equations:

$$\frac{d\mathbf{x}}{dt} = (\nabla^+ - \nabla^-)^T \cdot K \cdot \mathbf{x}^{\nabla^-} \tag{1}$$

with $K = diag(\kappa_1, \ldots, \kappa_r)$, and $\mathbf{x}^A$ the "vector-matrix" exponentiation: for $x = [x_1 \ldots x_q]^T \in \mathbb{R}^q$ and non-negative $A = [A_{ij}] \in \mathbb{R}^{p \times q}$, x^A denotes the element of $\mathbb{R}^p$ whose i^{th} component is $\prod_{j=1}^q x_j^{A_{ij}}$ (see Appendix A for an example).

This describes the continuous, deterministic model of a chemical reaction network, which is a limit of the stochastic model when all species are highly abundant [10]. One way to model CRNs is by using Petri Nets, as we recall below.

Definition 2. *A Petri Net is a tuple $N = <S, T, W, m_0>$, where S is a finite set of* places, *T is a finite set of* transitions, *$W : ((S \times T) \cup (T \times S)) \to \mathbb{N}$ is the arc weight mapping and $m_0 : P \to \mathbb{N}$ is the marking representing the initial distribution of tokens.*

A transition is enabled when all of its requirements are met (in the current marking, every place that has an incoming arc to the transition has at least as many tokens as the weight of its incoming arc), and it is fired by consuming all required tokens and producing new tokens.

Representing a CRN using Petri Nets is straightforward: places represent species (genes, proteins, complexes), and transitions represent reactions.

The most commonly used execution semantics of Petri Nets is the inter-leaving execution semantics: in each step, select one enabled transition non-deterministically, fire it, then repeat. This semantics describes totally asynchronous behaviour, which does not capture the concurrent nature of cellular behavior, where all reactions can happen in parallel. A better suited semantics, proposed in [7], and which we adapt in this paper, is presented below.

2.1   Max-Parallel Execution Semantics of Petri Nets

The max-parallel execution semantics can be informally described as "execute greedily as many transitions as possible in one step" [7]; we formalize this description in Definition 3.

The markings of a PN can be regarded as non-negative integer S-vectors. Its transition relation can be then described as a pair of $|S| \times |T|$ incidence matrices: ∇^- defined by $\forall s, t : \nabla^-(s, t) = W(s, t)$ and ∇^+ defined by $\forall s, t : \nabla^+(s, t) = W(t, s)$. Then their difference $\nabla = \nabla^+ - \nabla^-$, the composite change matrix, can be used to describe the reachable markings: for each sequence of transitions, w, $o(w)$ will denote T-the vector that maps every transition to its number of occurrences in w. Then, we have $reach(m_0) = \{m \mid \exists w : m = m_0 + \nabla \cdot o(w) \wedge$ w is a firing sequence[1] of $m_0\}$.

Definition 3. *A max-parallel execution step in a PN at state m is a positive T-vector v such that:*

1. *v is **compatible** with m (i.e., there are enough tokens to do everything, in any order): $0 \le m - \nabla^- v$*
2. *v is **exhaustive** (i.e., no reaction is enabled after firing): $\forall j \in T, m - \nabla^- v \not\ge r_j$, where r_j is the jth column of ∇^-.*

Figure 1 depicts the Petri Net of the CRN (we ignore the products):

$$3A + 2B \xrightarrow{\kappa_0} \cdots$$
$$5B + 3C \xrightarrow{\kappa_1} \cdots$$
$$C \xrightarrow{\kappa_2} \cdots$$

with initial marking $m_0 = (9, 9, 9)$, and its possible max-parallel strategies.

---

[1] A sequence of transitions that can be fired consecutively starting from a marking.

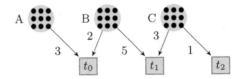

Fig. 1. A network with exactly 2 possible maximally parallel steps: $\{t_0 \times 3, t_2 \times 9\}$ and $\{t_0 \times 2, t_1, t_2 \times 6\}$

3   Piecewise Synchronous Execution Semantics of CRNs

In order to deal with resource allocation in CRNs, we construct an execution semantics of PNs that is piecewise synchronous (and includes the maximally parallel strategies): among all traces of execution of a PN, we single out a subset of semi-synchronous ones.

Execution proceeds in an alternation of resource allocation ("split") and depletion ("burst") (plus a phase of collection of the products between depletion and the next allocation). Allocation of tokens to their possible transitions is done via a $|T| \times |S|$ matrix α, where α_{ij} denotes the fraction of resource j being allocated to reaction i, meaning that:

$$\forall j \in S, \sum_{i \in T} \alpha_{ij} \leq 1 \tag{2}$$

The "burst" phase consists of the execution of all transitions in parallel until the available input are reduced to a small constant fraction of the initial amount (for reasons explained below). This small constant remainder we impose on our semantics is both the reason for the inequality sign in (2), and the reason our executions will be in the spirit of max-parallel executions, rather than max-parallel in the strict sense: whereas the max-parallel execution seeks to deplete all available resources, ours consumes them up to a fixed level (noted ϵ in the following).

3.1   Resource Allocation: Relation to Max-Parallel Execution

Suppose a CRN and assume $\alpha \in \mathbb{R}_+^{|T| \times |S|}$ a resource allocation matrix defined as above, $m \in \mathbb{R}^{|S| \times 1}$ a marking of the PN (i.e., a resource array) and v a (potentially max-parallel) reaction vector. We note that zero-order reactions ($\emptyset \to \ldots$) are not taken into account, as the question of resource allocation does not apply to them.

We define the operation $\star$ as:

Definition 4. $(\alpha \star m)_j \overset{\text{def}}{=} \min_{i \in S} \left(\frac{\alpha_{ji}}{\nabla_{ij}} \cdot m_i \right).$

Then Theorem 1 states that our execution semantics encompasses the max-parallel strategy (each max-parallel strategy is associated with a resource allocation matrix α, see Appendix B for proof).

Theorem 1. $\forall v$ *compatible with a resource array* m *(and potentially max-parallel),* $\exists \alpha$ *resource allocation matrix s.t.* $v = \alpha \star m$. *Furthermore, if the CRN is unary, there is unicity of* α.

For bimolecular reactions, α defined as in Appendix B is no longer the unique solution of $v = \alpha \star m$; intuitively, for a bimolecular reaction $r_k : A + B \rightarrow \ldots$, a different resource allocation matrix α' can be created by allocating to r_k whatever amount of species A is not allocated in α. The use of min in Definition 4 ensures that $v = \alpha' \star m$ (see Appendix C for an example).

3.2 Unary CRNs and Growth Rate

Consider a CRN comprised exclusively of unimolecular reactions. Then, for m an initial marking, ∇ the composite change matrix, and α a resource allocation matrix, the state of the system after one execution with the α split is given by the matrix $(I + \nabla \cdot \alpha) \cdot m$, with I the identity matrix. More generally, after k iterations of the "split-burst" execution with the same split α, the state of the system is:

$$D_\alpha^k \cdot m, \quad \text{with} \quad D_\alpha = I + \nabla \cdot \alpha \tag{3}$$

Let $\lambda_1 > \lambda_2 > \cdots$, the eigenvalues of D_α, and $E(\lambda_i)$ the eigenspace associated to each λ_i. If the initial marking vector can be decomposed as $m = \sum_i m_i$, with $m_i \in E(\lambda_i)$, then we can rewrite:

$$D_\alpha^k \cdot m = \lambda_1^k \cdot [m_1 + \sum_{i \geq 2} \left(\frac{\lambda_i}{\lambda_1}\right)^k \cdot m_i] \tag{4}$$

If $\lambda_1 < 1$, given (4), the system will eventually go extinct. Also, if $m_1 \in E(\lambda_1)$ is not unique, one has redundancy of growth (i.e., growth on multiple species/sources). We thus assume that $\lambda_1 > 1$, alongside uniqueness of m_1 and unidimensionality of eigenspaces.

Under these assumptions, as $\frac{\lambda_i}{\lambda_1} < 1$, for a big-enough k, the state of the system will converge to $\lambda_1^k \cdot m_1$, meaning that the growth rate of the system is given by λ_1, the biggest eigenvalue of D_α.

4 Depletion Time of Unary CRNs

Consider a unary reaction $S_i \rightarrow \ldots @k$; the time evolution of the concentration of species S_i is given by the ordinary differential equation $\frac{dS_i}{dt} = -k \cdot [S_i]$. Then at time t, the concentration of species S_i is: $S_i(t) = S_i(0) \cdot e^{-kt}$. Equivalently, the mean time of depletion of the reaction (i.e., bringing the level of species S_i to a specified amount $0 < s_i \leq S_i(0)$) is

$$\tau = k^{-1} \log \left(\frac{S_i(0)}{s_i}\right) \tag{5}$$

We note that s_i is a convention (the remainder of the reaction), or rather $\frac{S_i(0)}{s_i}$ is the relative amount we consume off the input (e.g., $s_i = 1\% \, S_i(0)$); the point being that we cannot deplete the whole amount of S_i, as that would take time $\tau = \infty$. Herein lies the main difference between our method and max-parallel execution, as mentioned in the beginning of Sect. 3.

Now suppose n unary reactions with the same input:

$$S_i \rightarrow \ldots @k_j; j \in N_i, |N_i| = n \tag{6}$$

We then allocate in parallel the S_i's between the n reactions, according to our execution semantics: $\forall j \in N_i$, reaction j receives $\alpha_{ji} \cdot S_i(0)$ input, and has a remainder of $s_{i,j}$. Then the depletion time of reaction j is:

$$\tau_j = k_j^{-1} \cdot \log\left(\alpha_{ji} \cdot \frac{S_i(0)}{s_{i,j}}\right) \tag{7}$$

4.1 Isochronicity and Iso-remainder Assumptions

In order to have a synchronous execution, we fix the same depletion time, τ for all reactions of the unary CRN. Then, from (7):

$$\alpha_{ji} = \beta^{k_j} \cdot \epsilon_{i,j} \tag{8}$$

where $\beta = e^\tau$ and $\epsilon_{i,j} = \frac{s_{i,j}}{S_i(0)}$. In this notation, $\epsilon_{i,j}$ is the remaining *percentage* (relative amount) of the total amount of S_i available in the beginning of the split round (i.e., $S_i(0)$), after reaction j is executed.

Furthermore, if we assume the same relative amount, s_i, remains after executing all n reactions, we have:

$$\forall j \in N_i, \alpha_{ji} = \epsilon_i \cdot \beta^{k_j}, \tag{9}$$

with $\epsilon_i = \frac{s_i}{S_i(0)}$.

Under these assumptions, the dynamics of the system in state m, for species S_i, is given by:

$$\Delta m(\tau) = \nabla \cdot (\alpha_{\cdot i} - \epsilon_{\cdot i}) \cdot m, \tag{10}$$

where $\alpha_{\cdot i}$ denotes the ith column of the resource allocation matrix α:

$$\forall j \in T, \alpha_{ji} = \begin{cases} \epsilon_i \cdot \beta^{k_j}, & \text{if } j \in N_i \\ 0, & \text{otherwise} \end{cases}, \quad \text{and} \quad \forall j \in T, \epsilon_{ji} = \begin{cases} \epsilon_i, & \text{if } j \in N_i \\ 0, & \text{otherwise} \end{cases}.$$

Then, from (2)[2] and (9), we have $\epsilon_i = \frac{1}{\sum_{j \in N_i} 1 + \beta^{k_j}}$ and:

$$\hat{\alpha}_{S_i}(\tau, \hat{k}_{S_i}) \overset{\text{def}}{=} [\alpha_{\cdot i} - \epsilon_{\cdot i}] = [\frac{e^{\tau \cdot k_j} - 1}{\sum_j 1 + e^{\tau \cdot k_j}}] \tag{11}$$

---

[2] The inequality of (2) is here explicitly expressed via the remainder ϵ: $\sum_{i \in T} \alpha_{ij} \leq 1$ is the same as $\sum_{i \in T}(\alpha_{ij} + \epsilon_j) = 1$.

(NB: $\left[\frac{e^{\tau \cdot k_j}-1}{\sum_j 1+e^{\tau \cdot k_j}}\right]$ actually denotes a T-vector that has 0 in the components representing reactions $j \notin N_i$).

Then, by injecting (11) into (10), one can easily observe that when $\tau \to 0$, $\Delta m(\tau)$ recreates the usual ODE system dynamics (as in formula 1):

$$\frac{\Delta m}{\tau} \approx \nabla \cdot [k_j] \cdot m \tag{12}$$

(NB: when $x \to 0$, $e^x \approx 1 + x$).

5 Applications

Based on formulae (4) and (11), our method can be interpreted either in a concrete way, as an approximation of the real system's dynamics (and be used for simulation purposes), or in an abstract way, as an alternative to flux balance analysis ("Synchronous Balanced Analysis").

5.1 Concrete Interpretation: Approximation of System Dynamics

As an approximation of the dynamics, under the unary/isochronous/iso-remainder assumptions, ours is a temporised discrete execution dynamics, that, when $\tau \to 0$, recreates the usual ODE dynamics. If we fix τ the execution time-step, and $\hat{k}$ the reaction rate vector, we can determine α, the resource allocation matrix, and ϵ, the remainder percentage (cf. (11)). The "iso" assumption represents a way of decoupling production and consumption in the chemical network, in the spirit of Karr's modular systems [12]; intuitively, it can be interpreted as: "in a parallel execution of a reaction set, there is no waiting for the slowest reaction to complete".

As a simulation method, it can be viewed as a big-step approximation of an integrator, resembling a deterministic τ-leaping [13].

5.2 Abstract Interpretation: Synchronous Balanced Analysis

Conversely, if the resource allocation matrix α is fixed, our execution semantics can be interpreted as an alternative to Flux Balance Analysis [11], in order to determine the limitations of a metabolic system; this is maybe the most important application of our method. FBA is a constraint-based approach that creates a solution space based on stoichiometric information, which impose constraints on the flow of metabolites through the network. The (flux) solution space can be further reduced via optimization with respect to a mathematical "objective function" representing a biological objective that is relevant to the problem being studied. In the case of predicting growth, the objective is biomass production, which is mathematically represented by an "objective function" Z that indicates how much each reaction contributes to the phenotype.

Our method uses λ, the growth rate, as the objective function. Cf. (4) maximizing λ means finding the resource allocation matrix α with the maximal biggest eigenvalue. Once this α is fixed, (11) can be used in order to constrain the reaction rates, $\hat{k}$, as well as the time-step τ, and the remainder ϵ. Hence, our method can be used as a technique of refuting models.

The advantages, when compared to FBA, are twofold: firstly, our method is applicable to growth systems (the implicit assumption of FBA is that the system has reached steady state), thus taking into account the real system kinetics; secondly, the idea of maximizing the biomass is preserved, but the invention of an "objective biomass function" is no longer needed, as it emerges directly from our method: *the growth rate*. The downside of our "Synchronous Balanced analysis" resides in the difficulty of maximizing the biggest eigenvalue of matrix D_α.

6   Conclusions and Future Work

In this paper, we propose a piecewise synchronous approximation of the dynamics of a (growth) chemical reaction network: a parallel execution semantics of Petri Nets, based on resource allocation. Our method can be interpreted either as an approximation of the real dynamics of the system, or as a constraint method similar to flux balance analysis, and has the advantage of being able to characterize the behavior of a cell using only one construction: the resource allocation matrix α. Consequently, one can eliminate the mechanistic details that deal with resource allocation, and replace them by an abstract vector (α). Furthermore, when compared to flux balance analysis, our method is applicable to growth systems.

Future work. Since the method presented in this paper constitutes work in progress, we plan to extend it into several directions. Firstly, we are interested in the extension of our method to binary reactions: the depletion time cannot be derived in the same way as for unary reactions. A way to potentially (but not completely) relieve this issue is by assuming that no significant change in the concentration of one of the two reactants is being caused by any other reaction during one execution step. We plan on looking into Michelis-Menten like reduction schemes to alleviate this problem. Once the issue of bimolecular reactions is solved, we can construct untrivial examples based on real-life CRNs, that will allow us to test the quality of our method.

Secondly, we plan to use our method to determine possible correlations between growth rate and different model parameters (such as reaction rate constants).

Last but not least, we would compare our method to the τ-leaping simulation, as well as the allocation method proposed by Karr [12], and further study the quality of our method.

Appendix A CRN Mass-action Kinetic Equations

Consider the following chemical reaction network:

$$S_1 + S_2 \xrightarrow{\kappa_1} 2S_2$$
$$S_2 \xrightarrow{\kappa_2} S_1$$

Then $S = \{S_1, S_2\}$, $\nabla^- = \begin{bmatrix} 1 & 1 \\ 0 & 1 \end{bmatrix}$, $\nabla^+ = \begin{bmatrix} 0 & 2 \\ 1 & 0 \end{bmatrix}$, $\mathbf{x}^{\nabla^-} = \begin{bmatrix} x_1 x_2 \\ x_2 \end{bmatrix}$ and $\frac{dx}{dt} =$

$\begin{bmatrix} -1 & 1 \\ 1 & -1 \end{bmatrix} \cdot \begin{bmatrix} \kappa_1 & 0 \\ 0 & \kappa_2 \end{bmatrix} \cdot \begin{bmatrix} x_1 x_2 \\ x_2 \end{bmatrix} = \begin{bmatrix} -\kappa_1 x_1 x_2 + \kappa_2 x_2 \\ \kappa_1 x_1 x_2 - \kappa_2 x_2 \end{bmatrix}$

Appendix B Proof of Theorem 1

Proof. Given a resource array m, consider v a (potentially max-parallel) vector compatible at m:

$$\nabla^- v \leq m \tag{13}$$

Then we can construct $\alpha \in \mathbb{R}_+^{|T| \times |S|}$:

$$\alpha_{ji} = \frac{\nabla_{ij}^- \cdot v_j}{m_i} \tag{14}$$

s.t.

$$(\alpha \star m)_j = \min_{i \in S} \left\{ \frac{\alpha_{ji}}{\nabla_{ij}^-} \cdot m_i \right\} = \min \left\{ \frac{\nabla_{ij}^- \cdot v_j}{m_i} \cdot \frac{m_i}{\nabla_{ij}^-} \mid \nabla_{ij}^- \neq 0 \right\} = v_j \tag{15}$$

Furthermore,

$$\forall j \in S, \sum_{i \in T} \alpha_{ij} = \frac{\sum_{i \in T} \nabla_{ji}^- \cdot v_i}{m_j} \overset{(13)}{\Longrightarrow} \forall j \in S, \sum_{i \in T} \alpha_{ij} \leq 1 \tag{16}$$

i.e. α is indeed a resource-allocation matrix.

If all reactions of the CRN are unimolecular, then:

$$\forall j \in T, \exists! i_j \in S \quad \text{s.t.} \quad \nabla_{i_j j}^- \neq 0 \implies \forall j \in T, (\alpha \star m)_j = \frac{\alpha_{j i_j}}{\nabla_{i_j j}^-} \cdot m_{i_j} \tag{17}$$

hence the uniqueness of α. □

Appendix C Non-uniqueness of α for Bimolecular Reactions

Example 1. (Based on Fig. 1.) $m = \begin{bmatrix} 9 \\ 9 \\ 9 \end{bmatrix}$, $\nabla^- = \begin{bmatrix} 3 & 0 & 0 \\ 2 & 5 & 0 \\ 0 & 3 & 1 \end{bmatrix}$, and $v = \begin{bmatrix} 2 \\ 1 \\ 6 \end{bmatrix}$, one of the 2 possible maximally parallel steps $(\{t_0 \times 2, t_1, t_2 \times 6\})$.

Then $\exists \alpha = \begin{bmatrix} \frac{6}{9} & \frac{4}{9} & 0 \\ 0 & \frac{5}{9} & \frac{3}{9} \\ 0 & 0 & \frac{6}{9} \end{bmatrix}$, defined as in (14), s.t. $\alpha \star m = \begin{bmatrix} \frac{6}{9} \cdot 9 \cdot \frac{1}{3} \wedge \frac{4}{9} \cdot 9 \cdot \frac{1}{2} \wedge \infty \\ \infty \wedge \frac{5}{9} \cdot 9 \cdot \frac{1}{5} \wedge \frac{3}{9} \cdot 9 \cdot \frac{1}{3} \\ \infty \wedge \infty \wedge \frac{6}{9} \cdot 9 \cdot \frac{1}{1} \end{bmatrix} =$

$\begin{bmatrix} 2 \\ 1 \\ 6 \end{bmatrix} = v$.

By re-allocating the excess of species A to the first reaction, we get
$\alpha' = \begin{bmatrix} 1 & \frac{4}{9} & 0 \\ 0 & \frac{5}{9} & \frac{3}{9} \\ 0 & 0 & \frac{6}{9} \end{bmatrix}$, a resource-allocation matrix that also verifies $\alpha' \star m =$
$\begin{bmatrix} 1 \cdot 9 \cdot \frac{1}{3} \wedge \frac{4}{9} \cdot 9 \cdot \frac{1}{2} \wedge \infty \\ \infty \wedge \frac{5}{9} \cdot 9 \cdot \frac{1}{5} \wedge \frac{3}{9} \cdot 9 \cdot \frac{1}{3} \\ \infty \wedge \infty \wedge \frac{6}{9} \cdot 9 \cdot \frac{1}{1} \end{bmatrix} = \begin{bmatrix} 2 \\ 1 \\ 6 \end{bmatrix} = v$ (non-uniqueness of α in the bimolecular case).

References

1. Garey, M.R., Johnson, D.S., Sethi, R.: The complexity of flowshop and jobshop scheduling. Math. Oper. Res. **1**, 117–129 (1976)
2. Sinnen, O.: Task Scheduling for Parallel Systems. Wiley-Interscience, Hoboken (2007)
3. Pugatch, R.: Greedy scheduling of cellular self-replication leads to optimal doubling times with a log-Frechet distribution. PNAS **112**(8), 2611–2616 (2015)
4. Weiße, A.Y., Oyarzún, D.A., Danos, V., Swain, P.S.: Mechanistic links between cellular trade-offs, gene expression, and growth. PNAS **112**(9), E1038–E1047 (2015)
5. Păun, G., Rozenberg, G.: A guide to membrane computing. Theor. Comput. Sci. **287**(1), 73–100 (2002). doi:10.1016/S0304-3975(02)00136-6
6. Lévy, J.-J.: Réductions correctes et optimales dans le lambda-calcul. Ph.D. thesis, Université Paris 7, January 1978
7. Krepska, E., Bonzanni, N., Feenstra, A., Fokkink, W., Kielmann, T., Bal, H., Heringa, J.: Design issues for qualitative modelling of biological cells with petri nets. In: Fisher, J. (ed.) FMSB 2008. LNCS (LNBI), vol. 5054, pp. 48–62. Springer, Heidelberg (2008)
8. Fisher, J., Henzinger, T.A., Mateescu, M., Piterman, N.: Bounded asynchrony: concurrency for modeling cell-cell interactions. In: Fisher, J. (ed.) FMSB 2008. LNCS (LNBI), vol. 5054, pp. 17–32. Springer, Heidelberg (2008)
9. Picard, V.: Réseaux de réactions: de l'analyse probabiliste à la réfutation. Ph.D. thesis, Université de Rennes 1, December 2015
10. Kurtz, T.G.: Limit theorems for sequences of jump Markov processes approximating ordinary differential processes. J. Appl. Probab. **8**(2), 344–355 (1971)
11. Orth, J.D., Thiele, I., Palsson, B.Ø.: What is flux balance analysis? Nat. Biotechnol. **28**(3), 245–248 (2010). http://doi.org/10.1038/nbt.1614
12. Karr, J.R., et al.: A whole-cell computational model predicts phenotype from genotype. Cell **150**(2), 389–401 (2012). http://dx.doi.org/10.1016/j.cell.2012.05.044
13. Gillespie, D.T.: Approximate accelerated stochastic simulation of chemically reacting systems. J. Chem. Phys. **115**(4), 1716 (2001). doi:10.1063/1.1378322

Discrete and Network Modelling

Verification of Temporal Properties of Neuronal Archetypes Modeled as Synchronous Reactive Systems

Elisabetta De Maria[1]([✉]), Alexandre Muzy[1], Daniel Gaffé[2],
Annie Ressouche[3], and Franck Grammont[4]

[1] Université Côte d'Azur, CNRS, I3S, Sophia Antipolis, France
{edemaria,muzy}@i3s.unice.fr
[2] Université Côte d'Azur, CNRS, LEAT, Sophia Antipolis, France
Daniel.GAFFE@unice.fr
[3] Université Côte d'Azur, Inria, Sophia Antipolis, France
annie.ressouche@inria.fr
[4] Université Côte d'Azur, CNRS, LJAD, Nice, France
grammont@unice.fr

Abstract. There exists many ways to connect two, three or more neurons together to form different graphs. We call archetypes only the graphs whose properties can be associated with specific classes of biologically relevant structures and behaviors. These archetypes are supposed to be the basis of typical instances of neuronal information processing. To model different representative archetypes and express their temporal properties, we use a synchronous programming language dedicated to reactive systems (Lustre). The properties are then automatically validated thanks to several model checkers supporting data types. The respective results are compared and depend on their underlying abstraction methods.

1 Introduction

Since a few years, the investigation of neuronal micro-circuits has become an emerging question in Neuroscience, notably in the perspective of their integration with neurocomputing approaches [15]. We call archetypes specific graphs of a few neurons with biologically relevant structures and behaviors. These archetypes correspond to elementary and fundamental elements of neuronal information processing. Several archetypes can be coupled to constitute the elementary bricks of bigger neuronal circuits in charge of specific functions. For instance, locomotive motion and other rhythmic behaviors are controlled by well-known specific neuronal circuits called Central Generator Pattern (CPG) [16]. These CPG have the capacity to generate oscillatory activities, at various regimes, thanks to some specific properties at the neuronal and circuit levels.

The goal of this work is to formally study the behavior of different representative archetypes. At this aim, we model the archetypes using a synchronous language for the description of reactive systems (Lustre). Each archetype (and

© Springer International Publishing AG 2016
E. Cinquemani and A. Donzé (Eds.): HSB 2016, LNBI 9957, pp. 97–112, 2016.
DOI: 10.1007/978-3-319-47151-8_7

corresponding assumed behavior in terms of neuronal information processing) is validated thanks to model checkers.

Different approaches have been proposed in the literature to model neural networks (Artificial Neural Networks [4], Spiking Neural Networks [13], etc.). In this paper we focus on Boolean Spiking Neural Networks where the neurons electrical properties are described via an integrate-and-fire model [7]. Notice that discrete modeling is well suited because neuronal activity, as with any recorded physical event, is only known through discrete recording (the recording sampling rate is usually set at a significantly higher resolution than the one of the recorded system, so that there is no loss of information). We describe neural networks as weighted directed graphs whose nodes represent neurons and whose edges stand for synaptic connections. At each time unit, all the neurons compute their membrane potential accounting not only for the current input signals but also for the ones received along a given temporal window. Each neuron can emit a spike according to the overtaking of a given threshold. Such a modeling is more sophisticated than the one proposed by McCulloch and Pitts in [17], where the behavior of a neural network is expressed in terms of propositional logic and the present activity of each neuron does not depend on past events.

Spiking neural networks can be considered as reactive systems: their inputs are physiological signals coming from input synapses, and their outputs represent the signals emitted in reaction. This class of systems fits well with the synchronous approach based on the notion of a *logical time*: time is considered as a sequence of logical discrete *instants*. An instant is a point in time where external input events can be observed, along with the internal events that are a consequence of the latter. The synchronous paradigm can be implemented using synchronous programming languages. In this approach we can model an activity according to a logical time framing: the activity is characterized by a set of events expected at each logical instant and by their expected consequences. A synchronous system evolves only at these instants and is "frozen" otherwise (nothing changes between instants). At each logical instant, all events are instantaneously broadcasted, if necessary, to all parts of the system whose instantaneous reaction to these events contributes to the global system state. Synchronous programming languages being initially dedicated to digital circuits, this neural implementation could be easily mapped into a physical one.

Each instant is triggered by input events (the core information completed with the internal state computed from instantaneous broadcast performed during the instant frame). As a consequence, inputs and resulting outputs all occur simultaneously. This (ideal) *synchrony hypothesis* is the main characteristics of the synchronous paradigm [9]. Another major feature is also that it supports concurrency through a deterministic parallel composition. The synchronous paradigm is now well established relying on a rigorous semantics and on tools for simulation and verification.

Several synchronous languages respect this synchronous paradigm. All these languages have a similar expressivity. However, we choose here Lustre [9] *synchronous language* to express neuron behaviors more easily. Lustre defines operator networks interconnected with data flows and it is particularly well suited to express neuron networks. Lustre respects the *synchrony* hypothesis which divides time into discrete instants. It is a data flow language offering two main advantages: (1) it is *functional* with no complex side effects, making it well adapted to formal verification and safe program transformation; also, reuse is made easier, which is an interesting feature for reliable programming concerns; (2) it is a *parallel* model, where any sequencing and synchronization depends on data dependencies. Moreover, the Lustre formalism is close to temporal logic and this allows the language to be used for both writing programs and expressing properties as observers. Hence, Lustre offers an original verification means to prove that, as long as the environment behaves properly (i.e., satisfies some *assumption*), the program satisfies a given *property*. If we consider only safety properties, both the assumption and the property can be expressed by some programs, called *synchronous observers* [10]. An observer of a safety property is a program, taking as inputs the inputs/outputs of the program under verification, and deciding (e.g., by emitting an alarm signal) at each instant whether the property is violated. Running in parallel with the program, an observer of the desired property and an observer of the assumption made about the environment have just to check that either the alarm signal is never emitted (property satisfied) or the alarm signal is emitted (property violated). This can be done by a simple traversal of the reachable states of the compound program.

There exists several model checkers for Lustre that are well suited to our purpose: *Lesar* [11], *Nbac* [12], *Luke* [1], *Rantanplan* [6] and *kind2* [8]. Verification with Lesar is performed on an abstract (finite) model of the program. Concretely, for purely logical systems the proof is complete, whereas in general (in particular when numerical values are involved) the proof can be only partial. Indeed, properties related to values depend on the abstraction performed by the tool. In our experiment, some properties can be validated with Lesar, but some others need powerful abstraction techniques. Hence, we use Lustre tools such as Nbac, Luke, Rantanplan and kind2. To perform abstractions, Lesar and NBac use convex polyhedra [14] representation of integers and reals. On the other hand, Luke is also another k-induction model checker, however it is based on propositional logic. Finally, Rantanplan and kind2 rely on SMT (Satisfiabitily Modulo Theories) based k-induction. kind2 has been specifically developed to prove safety properties of Lustre models, it combines several resolution engines and it turns out that it is the most powerful model checker used in this paper. This overall approach is used here to verify temporal properties of archetypes using model-checking techniques.

The paper is organized as follows. In Sect. 2 we present the computational model we adopt. In Sect. 3 we introduce the basic archetypes (series, series with multiple outputs, parallel composition, negative loop, inhibition of a behavior, contralateral inhibition) and we show how they can be modeled using Lustre. More precisely, we illustrate how the behavior of a single neuron can be encoded in a Lustre node and how two or more neurons can be connected to form a circuit. In Sect. 4 we express in Lustre important temporal properties concerning the described archetypes and we verify the satisfaction of these properties using the above-mentioned model checkers. Finally, Sect. 5 is devoted to a final discussion on the obtained results and on the future work. For a quick introduction to Lustre and the code of all the tested properties, the reader can refer to [5].

2   Synchronous Reactive Neuron Model

We refer here to *synchronous reactive systems* as systems reacting under the synchronous assumption, i.e., as computing their states and sending instantaneously their output events when receiving input events. Synchronous reactive systems can be conceived as an abstraction of digital circuits. Therefore, to fit electronic/computational discreteness and finiteness, some assumptions according to the synchronous paradigm will be introduced.

We describe here first the structure of a neuron network as a graph. The dynamics of usual leaky integrate-and-fire spiking networks is presented later.

Definition 1. *A network of neurons is a weighted directed graph (G, w), where $G = (N, A)$ is a directed graph with $N = \{1, 2, \ldots, n\}$ the set of neuron indexes and $A = \{(i, j) \mid i, j \in N\}$ the set of ordered pairs of neuron indexes (synapses), and $w : A \to \mathbb{R}$ is the synapse weight function[1].*

In a leaky integrate-and-fire neuron, the *membrane potential* of the neuron integrates the values of the action potentials received from its input neurons.

Definition 2. *A usual leaky integrate-and-fire model is a structure $LIF_i = (I_i, Y_i, S_i, T_i, \Delta_i, \Lambda_i)$, where $I_i = \mathbb{B}$ is the input alphabet; $Y_i = \mathbb{B}$ is the output alphabet; $S_i = \mathbb{R}$ is the set of states defined as the set of values of membrane potential; $T_i = \mathbb{R}_0^+ \cup \{+\infty\}$ is the time base; $\Delta_i : I_i^m \times S_i \times T_i \to S_i$ is the transition function defined as*

$$
p_i' = \Delta_i(x_{i_1}, \ldots, x_{i_m}, p_i, t_i) = \begin{cases} \Sigma_{j \in Pred(i)} w_{ji} x_j & \text{if } p_i \geq \tau_i \\ r_i(t_i) p_i + \Sigma_{j \in Pred(i)} w_{ji} x_j & \text{otherwise} \end{cases}
$$

where $Pred(i)$ is the set of m predecessors of neuron $i \in N$, x_j is the input of neuron $i \in N$ received from neuron $j \in Pred(i)$, $w_{ji} = w(j, i) \in \mathbb{R}$ is the synapse weight from neuron $j \in N$ to neuron $i \in N$, $r_i(t_i)$ is the remaining potential coefficient (a decreasing function in time, usually $r_i(t_i) = exp(-\alpha t_i)$, with α a positive constant), and $\tau_i \in \mathbb{R}_0^+$ is the firing threshold; and $\Lambda_i : S_i \to Y_i$ is the output function defined as $\Lambda_i(p_i) = y_i = \begin{cases} 1 & \text{if } p_i \geq \tau_i \\ 0 & \text{otherwise} \end{cases}$.

For each synapse $(j, i) \in A$ between a neuron $j \in N$ and a neuron $i \in N$, $y_i \in \mathbb{B}$ is the *output spike value* emitted by neuron i, and $x_j \in \mathbb{B}$ is the *input spike value* of neuron i received from neuron j. If the membrane potential p_i is above the threshold τ_i, at the next transition the output spike value is set to $y_i = 1$ and the membrane potential is reset to $p_i = 0$. When the remaining potential coefficient r_i is a constant equal to 1, there is *no leak*, all the potential received

---

[1] With $w : A \to \mathbb{Q}$ for *synchronous reactive neurons* as discussed later.

at last transition remains in the neuron soma. When $r_i(t_i) = 0$ for each t_i, all the potential received at last transition is lost (the model is then equivalent to McCulloch & Pitts' model [17]).

The usual leaky integrate-and-fire model presented in Definition 2 is not compatible with our synchronous reactive system assumption: both state and time sets are possibly infinite (cf. $r_i(t_i) = exp(-\alpha t_i)$, the exponentially decreasing function defined for $t_i \in [0, +\infty]$). We will show now how to approximate and limit potential values to fit the synchronous reactive system assumption.

Let us define, for each neuron, the remaining potential $r(t) = exp(-\alpha t)$ as r^e, where r is a constant (e.g., $r = exp(-\alpha)$ or $r \in [0, 1]$) and $e \in \mathbb{R}_0^+ \cup \{+\infty\}$ is the *time elapsed* until the current time $t \in \mathbb{R}_0^+ \cup \{+\infty\}$. The membrane potential can now be defined as a sum of input values and a power law of remaining potentials, leading to $p(t) = \Sigma_{e=0}^{+\infty} \Sigma_{j=1}^{m} r^e x_j (t - e)^2$, where the input value $x(t-e)$ and the potential $p(t)$ are functions depending on the current time t. The membrane potential *integrates* both current input values and what remains from previous inputs. As remaining input potentials decrease with time following a power law, inputs received a long time ago can nevertheless be neglected. Only remaining input potentials greater than a *threshold error* ϵ can be considered, i.e., $r^e \geq \epsilon$. Thus only elapsed times $e \leq \frac{ln(\epsilon)}{ln(r)}$ can be taken into account, where $\sigma = \frac{ln(\epsilon)}{ln(r)}$ is the *integration time window*, i.e., the period over which the neuron integrates past input values. For example, an error $\epsilon = 1\%$ and a remaining coefficient $r = 50\%$ correspond to an integration window $\sigma = 6.64$. This means that, *sliding the integration window* of a width equal to σ, at each time t, no input older that $e = 6.64$ will be considered, leading to an error of $\epsilon = 1\%$ in the membrane potential. The *time-dependence of the membrane potential is not anymore infinite but bounded to* $[t-\sigma, t]$. State changes need now to be finite. If we discretize each time step with $t, e \in \mathbb{N}_0$ (leading to $\sigma = \lceil \sigma \rceil = \lceil 6.64 \rceil = 7$ for the previous example), the membrane potential $p(t)$ consists of a sum $\Sigma_{j=1}^{m} x_j(t) + r \Sigma_{j=1}^{m} x_j(t-1) + r^2 \Sigma_{j=1}^{m} x_j(t-2) + r^3 \Sigma_{j=1}^{m} x_j(t-3) + \ldots + r^\sigma \Sigma_{j=1}^{m} x_j(t-\sigma)$, i.e., $p(t) = \Sigma_{e=0}^{\sigma} r^e \Sigma_{j=1}^{m} x_j(t-e)$.

Thanks to the previous time boundness and discreteness, each time t can be considered as a particular *transition*. The *computation of the membrane potential now depends on a finite memory of maximum size* $m \times (\sigma + 1)$, with $m, \sigma \in \mathbb{N}_0$.

A last simplification of the usual leaky integrate-and-fire model presented in Definition 2 concerns the real values of both membrane potential and synaptic weights. Indeed, real numbers are approximated by computers as floating-point values. However, rational numbers are needed to get efficient results from model checkers.

In our case, notice that this assumption well fits with remaining coefficients (that can easily be approximated by Taylor series or simple percentages).

---

[2] For the sake of simplicity, we assume that all the synaptic weights are equal to 1; supposing that neuron i has m predecessors, we write $\Sigma_{j=1}^{m} x_j$ to denote $\Sigma_{j \in Pred(i)} x_j$; when there is no ambiguity on the neuron index, we do not indicate it.

Accounting for all the previous assumptions on usual leaky integrate-and-fire neurons, the following definition can be provided for their mapping to synchronous reactive systems (finite state automata are proved to be equivalent to synchronous programs as Lustre or Esterel [3]).

Definition 3. *A synchronous reactive neuron* implements *a leaky integrate and fire model* as *a finite state machine (FSM)* $SRN_i = (I_i, Y_i, S_i, \Delta_i, \Lambda_i)$, *where $I_i = \mathbb{B}$ is the* input alphabet; $Y_i = \mathbb{B}$ *is the* output alphabet; $S_i = \mathbb{Q}$ *is the set of membrane potential values; $\Delta_i : I_i^{m \times (\sigma_i + 1)} \to S_i$ is the* transition function *defined as $p_{i_k} = \Delta_i(\mathbf{X_k}) = \mathbf{W} \mathbf{X_k} \mathbf{R}$, where*

- $\mathbf{W} = [w_1, w_2, \ldots, w_m]$ *is the* vector of synaptic weights *(each column corresponds to an input $j \in \{1, \ldots, m\}$ and $w_j \in \mathbb{Q}$),*

- $\mathbf{X_k} = \begin{bmatrix} x_{10} & x_{11} & \cdots & x_{1\sigma_i} \\ x_{20} & x_{21} & \cdots & x_{2\sigma_i} \\ \vdots & \vdots & \ddots & \vdots \\ x_{m0} & x_{m1} & \cdots & x_{m\sigma_i} \end{bmatrix}_k$ *is a matrix of Boolean stored input values (each row corresponds to an input $j \in \{1, \ldots, m\}$ and each column to an elapsed time $e_i \in \{0, \ldots, \sigma_i\}$ with $\sigma_i \in \mathbb{N}_0$),*

- $\mathbf{R} = \begin{bmatrix} 1 \\ r_i \\ \vdots \\ r_i^{\sigma_i} \end{bmatrix}$ *is the* vector of remaining coefficients *with $r_i \in \mathbb{Q}$, and*

- $\tau_i \in \mathbb{Q}$ *is the* firing threshold.

In particular $\mathbf{X_0} = \begin{bmatrix} I_1 & 0 & \cdots & 0 \\ I_2 & 0 & \cdots & 0 \\ \vdots & \vdots & \ddots & \vdots \\ I_{m0} & 0 & \cdots & 0 \end{bmatrix}$ *and* $\mathbf{X_k} = \begin{cases} shift([\mathbf{0}], I_k) & if\ p_{i_{k-1}} \geq \tau_i \\ shift(\mathbf{X_{k-1}}, I_k) & otherwise \end{cases}$

where shift is a shifting matrix function: $M^{m \times (\sigma_i + 1)} \times V^m \to M^{m \times (\sigma_i + 1)}$

$shift(M, V) = \begin{bmatrix} V_1 & M_{10} & \cdots & M_{1(\sigma_i - 1)} \\ V_2 & M_{20} & \cdots & M_{2(\sigma_i - 1)} \\ \vdots & \vdots & \ddots & \vdots \\ V_m & M_{m0} & \cdots & M_{m(\sigma_i - 1)} \end{bmatrix}.^3$

Lastly, $\Lambda_i : S_i \to Y_i$ is the output function, *with $\Lambda_i(p_{i_k}) = \begin{cases} 1 & if\ p_{i_{k-1}} \geq \tau_i \\ 0 & otherwise. \end{cases}$*

This formalization allows for the characterization of each neuron through a parameter triplet $(\tau_i, r_i, \sigma_i) \in \mathbb{Q} \times \mathbb{Q} \times \mathbb{N}_0$, i.e., the firing threshold τ_i, the remaining coefficient r_i, and the integration window σ_i.

3   Encoding Neuronal Archetypes in Lustre

The basic archetypes we take into account are the following ones (see Fig. 1).

---

[3] It is not a classical finite state machine because the matrix $\mathbf{X_k}$ considers the current inputs as well as the past inputs received within the integration time window.

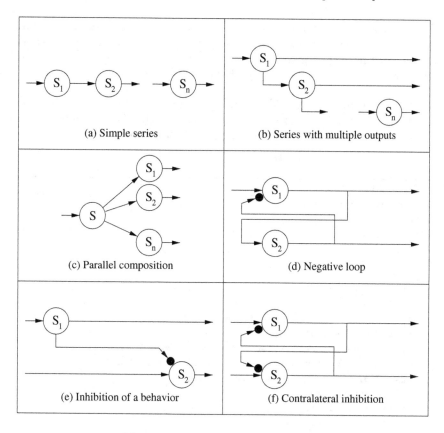

(a) Simple series

(b) Series with multiple outputs

(c) Parallel composition

(d) Negative loop

(e) Inhibition of a behavior

(f) Contralateral inhibition

Fig. 1. The basic neuronal archetypes

- **Simple series.** It is a sequence of neurons where each element of the chain receives as input the output of the preceding one. The input (resp. output) of the first (resp. last) neuron is the input (resp. output) of the network. The spike emission of each neuron is constrained by the one of the preceding neuron.
- **Series with multiple outputs.** It is a series where, at each time unit, we are interested in knowing the outputs of all the neurons (i.e., all the neurons are considered as output ones).
- **Parallel composition.** There is a set of neurons receiving as input the output of a given neuron. All neurons working in parallel are considered as output ones.
- **Negative loop.** It is a loop consisting of two neurons: the first neuron activates the second one while the latter inhibits the former one. The inhibited neuron is supposed to oscillate.
- **Inhibition of a behavior.** There are two neurons, the first one inhibiting the second one. After a certain delay, the first neuron is supposed to be activated and the second one to be inhibited.

- **Contralateral inhibition.** There are two or more neurons, each one inhibiting the other ones. The expected behavior is of the kind "winner takes all", that is, starting from a given time only one neuron becomes (and stays) activated and all the other ones are inhibited.

In the following we provide a Lustre implementation of neurons and archetypes. A Boolean neuron with one predecessor (that is, one input neuron), can be modeled as the Lustre node described in Program 1.

Program 1 Basic neuron node.

```
node neuron105 (X:bool) returns(S:bool);    var
    V:int;
    threshold:int;
    w:int;
    rvector: int^5;
    mem:int^1*5;
    localS: bool;
  let
    w=10; threshold=105; rvector=[10,5,3,2,1];
    mem[0]=if X then w else 0;
    mem[1..4]=0^4->if pre(S) then 0^4 else pre(mem[0..3]);
    V=mem[0]*rvector[0]+mem[1]*rvector[1]+mem[2]*rvector[2]
        +mem[3]*rvector[3]+mem[4]*rvector[4];
    localS=(V>=threshold);
    S= false -> pre(localS);
  tel
```

In the node **neuron105** (where the firing threshold is set to 105), X is the Boolean flow representing the input signal of the neuron, w is the synaptic weight of the input edge, rvector is the vector containing the different values the remaining coefficient can take along the integration window (from the biggest to the smallest one), and the vector mem keeps trace of the received signals (from the current one to the one received at the time $t - \sigma$)[4]. More precisely, at each time unit the first column of vector mem contains the current input (multiplied by the synaptic weight of the input edge) and, for each i greater than 0, the value of the column i is defined as follows: (i) it equals 0 at the first time unit (initialization) and (ii) for all following time units it is reset to 0 in case of spike emission at the preceding time unit and it takes the previous time unit value of the column $i - 1$ otherwise. Variable localS is used to introduce a delay in the spike emission.

---

[4] Observe that all the parameters are multiplied by 10 in order to only deal with integer numbers (and thus to be able to use all the model checkers available to Lustre).

The generalization to a node with m predecessors is straightforward. Thanks to the modularity of Lustre, archetypes can be easily encoded starting from basic neurons. As an example, a simple series composed of three neurons of type `neuron105` is described in Program 2.

Program 2 Simple series of three neurons.

```
node series3 (X:bool) returns(S:bool);
  var
    chain:bool^3;
  let
    chain[0]=neuron105(X);
    chain[1..2]=neuron105(chain[0..1]);
    S=chain[2];
  tel
```

In the node `series3`, each position of the vector `chain` refers to a different neuron of the chain. As far as the first neuron is concerned, it is enough to call the node `neuron105` with the input of the series as input. For the other neurons of the chain, their behavior is modeled by calling `neuron105` with the output of the preceding neuron as input. The output of the node is the one of the last neuron of the series.

4   Encoding and Verifying Temporal Properties of Archetypes in Lustre

The behavior of each archetype can be validated thanks to the use of model checkers such as Lesar, Nbac, Luke, Rantanplan, and kind2 (the last four ones have been used to deal with some properties involving integer constraints Lesar is not able to treat). For each archetype, one or two properties have been encoded as Lustre nodes and tested on some instances of the archetype. To illustrate our purpose, we show the encoding of the first two properties (the implementation of the other properties can be found in [5]). Most of the properties are tested here for all possible inputs and one or more set(s) of parameters for the given archetype.

4.1   Simple Series (see Fig. 1(a))

Given two series with the same triplets of parameters $(\tau, r, \sigma) \in \mathbb{Q} \times \mathbb{Q} \times \mathbb{N}_0$ and the same synaptic weights, the first series being shorter than the second one, we want to check whether the first series is always in advance with respect to the second one. More precisely, the property we test is the following one:

Property 1 *(Comparison of series with same parameters). Given two series with the same neuron parameters and different length (i.e., with a different number of neurons), at each step, the number of spikes emitted by the shorter series is greater or equal than the number of spikes emitted by the longer one.*

The node `prop1` (described in Program 3) expresses an observer of Property 1 in Lustre.

Program 3 Observer of Property 1

```
node prop1(X:bool) returns(S:bool);
  var
    A1,A2:bool;
    C1,C2;
  let
    A1=seriesA_sp(X);
    A2=seriesB_sp(X);
    C1=bool2int(A1)->if A1 then pre(C1)+1 else pre(C1);
    C2=bool2int(A2)->if A2 then pre(C2)+1 else pre(C2);
    S=(C1-C2)>=0;
  tel
```

Let `seriesA_sp` (resp. `seriesB_sp`) be the Lustre node encoding the first (resp. second) series (corresponding neurons in the two series have the same parameter triplets). In the node `prop1`, `C1` (resp. `C2`) keeps trace of the number of spikes emitted by the first (resp. second) series until the current time unit. The model checkers Lesar, Nbac, Luke, Rantanplan and kind2 verify whether, *whatever is the value of the input flow variable X* (which is common to the two series), the property is true, that is, `C1` is greater or equal than `C2`.

Another interesting property concerning simple series is the following one:

Property 2 *(Comparison of series with different parameters). Given two series with different neuron parameters and different length, they always have the same behavior.*

The node `prop2` (described in Program 4) encodes such a property in Lustre.

Program 4 Observer of Property 2

```
node prop2 (X:bool) returns(S:bool);
  var
    s1,s2:bool;
  let
    s1=seriesA_dp(X);
    s2=seriesB_dp(X);
    S=(s1=s2);
  tel
```

At each step, the output of the node is true if the output of the two series `seriesA_dp` and `seriesB_dp` is the same (provided that they receive the same input flow X). Such a property can be exploited in order to reduce a given neural network (if a given series has exactly the same behavior than a shorter one, it can be replaced by the second one). As an example, we found a series of 3 neurons showing the same behavior than a series of length 4 (neurons in the two series have different firing thresholds and synaptic weights).

4.2 Series with Multiple Outputs (see Fig. 1(b))

When dealing with a series with multiple outputs, we are interested in checking whether, soon or later, all the neurons of the sequence are able to emit a spike. It may not be the case if the parameters are not well chosen (for example, if the threshold of the first neuron is too high). The corresponding property formalized here is the following one:

Property 3 *(Firing ability in a series). Given a series with multiple outputs where the different neurons can have different parameters, there exists a time unit such that all the neurons have emitted.*

Let `prop3` be the node encoding the observer of Property 3. The output of `prop3` becomes (and stays) true after all the neurons of the series have emitted at least one spike. As an example of property violation, we have found a series of length 4 where, even if a flow of 1 (encoded as *true* in Lustre) is given as input, the last neuron is never able to emit. Observe that, given a series where all the neurons are able to emit, `prop3` only becomes true when the last neuron of the series emits a spike. In order to force the property to be immediately true, it is possible to take advantage of the node `always_since` from Lustre distribution library.

4.3 Parallel Composition (see Fig. 1(c))

We are interested in knowing a lower and an upper bound to the number of neurons that can emit a spike at each time unit. The lower (resp. upper) bound is not necessarily 0 (resp. the number of parallel neurons). More precisely, the property we test is the following one:

Property 4 *(Lower/upper firing bounds in a parallel composition). Given a parallel composition of neurons, all with the same parameters, at each time unit, the number of emitted spike is in between a given interval.*

Let the output variable of the node encoding the parallel composition represent the global number of spikes emitted at each time unit by all the neurons in parallel. The observer of Property 4 checks whether the number of emitted spikes is always in between a lower bound and an upper bound. As an example, we have found a parallel composition of 3 neurons where the number of emitted spikes is always strictly lower than 3 (more precisely, it is always in between 0 and 2). This is due to the fact that, for one of the parallel neurons, the synaptic weight of the corresponding input edge is too low and it is never able to emit.

4.4 Negative Loop (see Fig. 1(d))

In this case the inhibited neuron is expected to oscillate. In Property 5 we express an oscillation with a period of two time units.

Property 5 *(Oscillation in a negative loop). Given a negative loop (where the two neurons do not necessarily have the same parameters), the inhibited neuron oscillates with a pattern of the form* false, false, true, true.

Let Out be the output of the inhibited neuron of the Lustre node encoding the negative loop archetype. In the observer of Property 5, we check that (i) if Out is true, then it was false two time units ago and (ii) if Out is false then it was true two time units ago. For several parameters, if we inject only 1 as input of the archetype, such a property is satisfied. Observe that, to test the property satisfaction under some specific conditions, e.g., when the input variable X is equal to true, it is sufficient to introduce a new output variable SS defined as the disjunction of the current output variable S and the negation of the condition. (e.g., SS=S or X=false).

4.5 Inhibition of a Behavior (see Fig. 1(e))

To validate this archetype we need to verify that, at a certain instant, the inhibited neuron stops emitting spikes. In particular, the property we encoded is the following one:

Property 6 *(Fixed point inhibition). Given an inhibition archetype (where the two neurons do not necessarily have the same parameters), at a certain time the inhibited neuron can only emit* false *values.*

The output of the observer of Property 6 is true if the variable representing the output of the inhibited neuron of the archetype cannot pass from false to true. For appropriate parameter values, if we inject only 1 values, such a property turns out to be true.

4.6 Contralateral Inhibition (see Fig. 1(f))

For such an archetype the expected behavior is of the kind "winner takes all", that is, at a certain point only one neuron is activated and the other ones cannot emit spikes.

Property 7 *(Winner takes all in a contralateral inhibition). Given a contralateral inhibition archetype with two neurons (where the two neurons do not necessarily have the same parameters), at a given time, one neuron is activated and the other one is inhibited.*

In the observer of Property 7 we test whether, at each time unit, one neuron is activated and the other one inhibited. Such a property turns out to be true for several parameters (if only 1 values are injected). Let $w2$ (resp. $w4$) be the synaptic weight of the inhibiting input edge of the first (resp. second) node. In

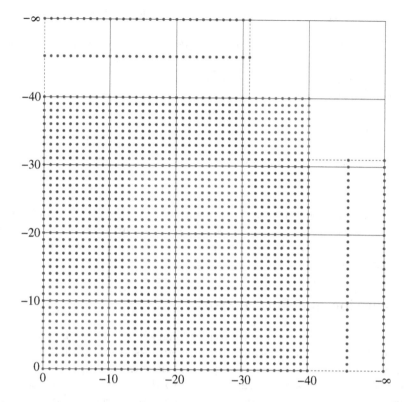

Fig. 2. Verification of prop7 for the different values of $(w2, w4)$ (Color figure online)

Fig. 2, blue points represent the pairs $(w2, w4)$ for which the property is verified starting from a time unit lower than or equal to 4 and red points are associated to the pairs for which the property is not verified within 10 time units (for some fixed parameter triplets).

4.7 Comparison of the Model Checkers

A synthesis of the outputs of the five model checkers is summarized for each property in Table 1:

Notice that, when a model checker gives a negative answer, it does not necessarily mean that the property is false; it can be an indication of the fact that the model checker is not able to conclude. In this experiment, Lesar has difficulties to handle complex integer constraints. Nbac goes further but it is quickly stopped by the polyhedra approach. Luke and its extension Rantanplan give similar results with sometimes a very long computation time. kind2 works quickly and it is able to prove more general properties than Luke and Rantanplan. For instance, Luke and Rantanplan allow for the identification of the pair of weights which stabilize the "Contralateral inhibition" (see Fig. 2) while kind2 is able to

Table 1. Comparison of the five model checkers

	Lesar	Nbac	Luke	Rantanplan	Kind2
Simple series (prop1)	No	Yes	very long time!	Yes	Yes
Simple series (prop2)	No	exit before!	Yes	very long time!	Yes
Series with multiple outputs	No	exit before!	Yes	Yes	Yes
Parallel composition	No	exit before!	Yes	Yes	Yes
Negative loop	No	exit before!	Yes	Yes	Yes
Inhibition of a behavior	No	Yes	Yes	Yes	Yes
Contralateral inhibition	No	Yes	Yes	Yes	Yes

straightly give us an infinite set of pair solutions. For the sake of completeness, we also tested the nuXmv model checker [2] but perhaps we could not find the good abstraction (neither too coarse, nor too thorough), so we could not get satisfying results.

5   Discussion and Future Work

In this work, we show how the synchronous language Lustre can be an effective tool to model, specify, and verify neuronal networks. More precisely, we illustrate how some basic neuronal archetypes and their expected properties can be encoded as Lustre nodes and verified thanks to the use of model checkers. For each archetype, we propose one or two representative properties that have been identified after deep discussions with neurophysiologists and, in particular, with the last author of this paper. As a first future work, we intend to propose a more general version of some properties (e.g., expressing oscillation without exactly knowing its period).

We choose to use Lustre because its declarative syntax is more adapted to our class of problems than an imperative language such as Esterel and because several model checkers integrating the symbolic manipulation of integer constraints are at Lustre user's disposition. However, these motivations do not prevent us from considering to use Light-Esterel in the future; the third and fourth author of this work are actually working on extending the expressivity of the declarative part of this language and developing a dedicated model checker. Particularly, this new model checker should integrate a new way to characterize and verify properties based on Linear Decision Diagram (LDD). This representation would allow to identify input parameter intervals of values for which a property holds.

As far as we know, this work constitutes the first attempt to automatically verify the temporal properties of fundamental neuronal archetypes in terms of neuronal information processing (e.g. a negative loop with certain parameters presents a certain oscillating behavior). From there, we will now be able to apply this new approach to all the possible archetypes of 2, 3 or more neurons, up to falling on archetypes of archetypes. One of the questions to ask then will be: are the properties of these archetypes of archetypes simply an addition of the individual constituting archetypes properties or something more? Another one will be: can we understand the computational properties of large ensembles of neurons simply as the coupling of the properties of individual archetypes, as it is for the alphabet and words, or is there something more again?

Acknowledgements. The authors would like to thank Gérard Berry for an inspiring talk at the *Collège de France* (concerning the checking of temporal properties of neuronal structures) as well as for having indicated us the researchers competent at the use of synchronous programming language libraries (in Sophia Antipolis).

References

1. Luke webpage. http://www.it.uu.se/edu/course/homepage/pins/vt11/lustre
2. Nuxmv webpage. https://nuxmv.fbk.eu/
3. Berry, G., Cosserat, L.: The ESTEREL synchronous programming language and its mathematical semantics. In: Brookes, S.D., Roscoe, A.W., Winskel, G. (eds.) CONCURRENCY 1984. LNCS, vol. 197, pp. 389–448. Springer, Heidelberg (1985). doi:10.1007/3-540-15670-4_19
4. Das, S.: Elements of artificial neural networks [book reviews]. IEEE Trans. Neural Netw. **9**(1), 234–235 (1998)
5. De Maria, E., Muzy, A., Gaffé, D., Ressouche, A., Grammont, F.: Verification of Temporal Properties of Neuronal Archetypes Using Synchronous Models. Research report 8937, UCA, Inria; UCA, I3S; UCA, LEAT; UCA, LJAD, July 2016. https://hal.inria.fr/hal-01349019
6. Franzén, A.: Using satisfiability modulo theories for inductive verification of lustre programs. Electron. Notes Theor. Comput. Sci. **144**(1), 19–33 (2006)
7. Gerstner, W., Kistler, W.: Spiking Neuron Models: An Introduction. Cambridge University Press, New York (2002)
8. Hagen, G., Tinelli, C.: Scaling up the formal verification of lustre programs with SMT-based techniques. In: Formal Methods in Computer-Aided Design, FMCAD 2008, Portland, Oregon, USA, pp. 1–9, 17–20 November 2008
9. Halbwachs, N.: Synchronous Programming of Reactive Systems. Kluwer Academic, Dordrecht (1993)
10. Halbwachs, N., Lagnier, F., Raymond, P.: Synchronous observers and the verification of reactive systems. In: Nivat, M., Rattray, C., Rus, T., Scollo, G. (eds.) Algebraic Methodology and Software Technology (AMAST '93). Workshops in Computing, pp. 83–96. Springer, London (1994)
11. Halbwachs, N., Raymond, P.: Validation of synchronous reactive systems: from formal verification to automatic testing. In: Thiagarajan, P.S., Yap, R.H.C. (eds.) ASIAN 1999. LNCS, vol. 1742, p. 1. Springer, Heidelberg (1999)
12. Jeannet, B.: Dynamic partitioning in linear relation analysis: application to the verification of reactive systems. Formal Methods Syst. Des. **23**(1), 5–37 (2003)
13. Maass, W., Graz, T.U.: Lower bounds for the computational power of networks of spiking neurons. Neural Comput. **8**, 1–40 (1995)
14. Maréchal, A., Fouilhé, A., King, T., Monniaux, D., Périn, M.: Polyhedral approximation of multivariate polynomials using Handelman's theorem. In: Jobstmann, B., Leino, K.R.M. (eds.) Verification, Model Checking, and Abstract Interpretation. LNCS, vol. 9583, pp. 166–184. Springer, Heidelberg (2016)

15. Markram, H.: The blue brain project. Nat. Rev. Neurosci. **7**(2), 153–160 (2006)
16. Matsuoka, K.: Mechanisms of frequency and pattern control in the neural rhythm generators. Biol. Cybern. **56**(5–6), 345–353 (1987)
17. McCulloch, W.S., Pitts, W.: A logical calculus of the ideas immanent in nervous activity. Bull. Math. Biophys. **5**(4), 115–133 (1943)

Relationship Between the Reprogramming Determinants of Boolean Networks and Their Interaction Graph

Hugues Mandon[1(✉)], Stefan Haar[2], and Loïc Paulevé[1]

[1] LRI UMR 8623, Univ. Paris-Sud – CNRS,
Université Paris-Saclay, Orsay, France
hugues.mandon@laposte.net
[2] LSV, ENS Cachan, Inria, CNRS, Université Paris-Saclay, Cachan, France

Abstract. In this paper, we address the formal characterization of targets triggering cellular trans-differentiation in the scope of Boolean networks with asynchronous dynamics. Given two fixed points of a Boolean network, we are interested in all the combinations of mutations which allow to switch from one fixed point to the other, either possibly, or inevitably. In the case of existential reachability, we prove that the set of nodes to (permanently) flip are only and necessarily in certain connected components of the interaction graph. In the case of inevitable reachability, we provide an algorithm to identify a subset of possible solutions.

1 Introduction

In the field of regenerative medicine, an emerging way to treat patients is to reprogram cells, leading, for instance, to tissue or neuron regeneration. Such a challenge has become realistic after first experiments have shown that some of the cell fate decisions can be reversed [15]. Whereas the cells go through several multipotent states before reaching a differentiated state, the differentiation process can be inversed, producing induced pluripotent stem cells (iPSCs) from an already differentiated cell. By using a distinct differentiation path, this allows to "transform" the type of a cell. Alternatively, it is also possible to directly perform a trans-differentiation without necessarily going (back) through a multipotent state [7,9].

In the aforementioned work, the de- and trans-differentiation has been achieved by targeting specific genes, that we refer to as *Reprogramming Determinants* (RDs), through the mediation of their transcription factors [6,15].

The computational prediction of RDs requires to assess multiple features of the cell dynamics and the reprogramming strategy, such as the impact of the kind of perturbations (persistent versus temporary) and of their order; the nature of targeted cell type (differentiated/pluripotent), and the desired inevitability of their reachability (fidelity); the nature and duration of the triggered cascade of regulations (efficiency); and finally, the RD robustness with respect to initial state heterogeneity among cell population, and with respect to uncertainties in the computational model.

© Springer International Publishing AG 2016
E. Cinquemani and A. Donzé (Eds.): HSB 2016, LNBI 9957, pp. 113–127, 2016.
DOI: 10.1007/978-3-319-47151-8_8

So far, no general framework allows to efficiently encompass those features to systematically predict best combinations of RDs in distinct cellular reprogramming events.

In this paper, we address the identification of RDs from *Boolean Networks* (BNs) which model the dynamics of gene regulation and signalling networks. The state of the components (or nodes) of the networks are represented by Boolean variables, and the state changes are specified by Boolean functions which associate the next state of nodes, given the (binary) state of their regulators [2,16]. BNs are well suited for an automatic reasoning on large biological networks where the available knowledge is mostly about activation and inhibition relations [1]. Such activation/inhibition relations between components form a signed directed graph, that we refer to as the *Interaction Graph*.

In this work, we make the assumption that the differentiated cellular states correspond to the *attractors* of the dynamics of the computational model, i.e., the long-run behaviours. In the scope of BNs, those attractors can be of two kinds: either a single state (referred to as a fixed point), or a terminal cyclic behaviour.

The relationship between the IG of BNs and the number of their attractor has been extensively studied [2,13,14]. However, little work exists on the characterization of the perturbations which trigger a change of attractor. Currently, most of RDs prediction are performed using statistical analysis on expression data in order to rank candidate transcription factors [3,10,12]. Whereas based on network models, those approaches do not allow to derive a complete set of solution for the reprogramming problem. In [6], the authors developed a heuristic to derive candidate RDs from a pure topological analysis of the interaction graph: the RDs are selected only in positive cycles that have different values in the started and target fixed points. However, there is no guarantee that the derived RDs can actually lead to a change of attractor in the asynchronous dynamics of the Boolean networks, and neither that the target fixed point is the only one reachable. Finally, [8] gives a formal characterization of RDs subject to temporal mutations which trigger a change of attractor in the synchronous semantics of conjunctive Boolean networks.

Contribution: This work relies on model checking and reachability analysis, that have been proved useful and successful in previous studies [1,11].

Given a BN, all of whose attractors are fixed points, given an initial fixed point and a target fixed point, we provide a characterization of the candidate RDs (set of nodes) with respect to the interaction graph and for two settings of cellular reprogramming:

- with a permanent perturbation of RDs, the target fixed point becomes reachable in the asynchronous dynamics of the BN;
- with a permanent perturbation of RDs, the target fixed point is the sole reachable attractor in the asynchronous dynamics of the BN.

For the first case, we prove that all the RDs are distributed among particular strongly connected components of the interaction graph, and we give algorithms to determine them in both settings. In the second case, we prove that only some of them are distributed among strongly connected components of the interaction graph. We provide an algorithm to identify possible combination of permanent perturbations leading to inevitable reachability of the target fixed point. Whereas the algorithm may miss some solutions, all returned solutions are correct.

Outline: Section 2 gives the definitions and basic properties of BNs and of their asynchronous dynamics. The formalization of the BN reprogramming problem with permanent perturbations of nodes is established in Sect. 3. Section 4 states the main results on the characterization of RDs with respect to the interaction graph of BNs. An algorithm to enumerate all RDs by exploiting this characterization is given in Sect. 5. Finally, Sect. 6 discusses the results and sketches future work.

Notations

Given a finite set I, 2^I is the power set of I, $|I|$ the cardinality. Given a positive integer n, $[n] = \{1, \ldots, n\}$.

Given a Boolean state $x \in \{0, 1\}^n$ and set of indexes $I \subset [n]$, $\bar{x}^I$ is the state where $\bar{x}_i^I = x_i$ if $i \notin I$ and $\bar{x}_i^I = 1 - x_i$ if $i \in I$. Similarly, given $x, y \in \{0, 1\}^n$, $x_{[x_I = y_I]}$ denotes the state where for all $i \in I$, $(x_{[x_I = y_I]})_i = y_i$ and for all $i \notin I$, $(x_{[x_I = y_I]})_i = x_i$.

2 Background

In this section, we give the formal definition of Boolean networks, their interaction graph and transition graph in the asynchronous semantics. Finally, we recall the main link between their attractors and the positive cycles in their interaction graph.

2.1 Definitions

Boolean Network (BN): A BN is a finite set of Boolean variables, each of them having a Boolean function. This function is a logical Boolean function depending from the network's variables and determining the next state of the variable.

Definition 1 (Boolean Network (BN)). *A Boolean Network is a function f such that:*

$$f : \qquad \{0, 1\}^n \to \{0, 1\}^n$$
$$x = (x_1, \ldots, x_n) \mapsto f(x) = (f_1(x), \ldots, f_n(x))$$

Example 1. An example of BN of dimension 3 $(n = 3)$ is

$$f_1(x) = x_3 \vee (\neg x_1 \wedge x_2)$$
$$f_2(x) = \neg x_1 \vee x_2$$
$$f_3(x) = x_3 \vee (x_1 \wedge \neg x_2)$$

Interaction Graph: To determine the RDs, we rely on a simplification of the interactions between the genes, and of the concentrations. A gene will either be active or inhibited. Gene interactions are simplified likewise, a gene either activates or inhibits another gene, and we ignore time scales. With this in mind, an *interaction graph* (Definition 2) can be build: genes are the vertices, and the interactions are the oriented arcs, labelled either $+$ or $-$, if it is an activation or an inhibition.

Definition 2 (Interaction Graph). *An interaction graph is noted as $G = (V, E)$, with V being the vertex set, and E being the directed, signed edge set, $E \subset (V \times V \times \{-, +\})$.*

A cycle between a set of nodes $C \subseteq V$ is said positive (resp. negative) if and only if there is an even (odd) number of negative edges between those nodes.

An interaction graph can also be defined as an abstraction of a Boolean network: the functions are not given and not always known, but if a vertex u is used in the function f_v, there is an edge from u to v, negative if $f_v(x)$ contains $\neg x_u$ and positive if it contains x_u.

Definition 3 (Interaction Graph of a Boolean network $(G(f))$). *An interaction graph can be obtained from the Boolean network f: the vertex set is $[n]$, and for all $u, v \in [n]$ there is a positive (resp. negative) arc from u to v if $f_{vu}(x)$ is positive (resp. negative) for at least one $x \in \{0, 1\}^n$ (For every $u, v \in \{1, ..., n\}$, the function f_{vu} is the discrete derivative of f_v considering u, defined on $\{0, 1\}^n$ by: $f_{vu}(x) := f_v(x_1, .., x_{u-1}, 1, x_{u+1}, .., x_n) - f_v(x_1, .., x_{u-1}, 0, x_{u+1}, .., x_n)$).*

Given an interaction graph $G = (V, E)$, and one of its vertex $u \in V$, P_u denotes the set of ancestors of u, i.e., the vertices v for which there exists a path in E from v to u. Similarly, p_u is the set of the parents of u, i.e., $v \in p_u \Rightarrow (v, u, s) \in E$. Furthermore, $G[P_u]$ is the induced subgraph of G with P_u as vertex set.

Figure 1 gives an example of an interaction graph, which is also equal to $G(f)$, where f is the Boolean network of Example 1.

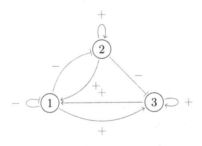

Fig. 1. Interaction graph of Example 1 A "normal" blue arrow means an activation, and a "flattened" red arrow means an inhibition

Transition Graph: We model the dynamics of a Boolean network f by *transitions* between its states $x \in \{0,1\}^n$. In the scope of this paper, we consider the *asynchronous semantics* of Boolean networks: a transition updates the value of only one vertex $u \in [n]$. From a single $x \in \{0,1\}^n$, one has different transitions for each vertex u such that $f_u(x) \neq x_u$. This leads to the definition of the *transition graph* (Definition 4) where vertices are all the possible states $\{0,1\}^n$, and edges correspond to asynchronous transitions.

Definition 4 (Transition graph). *The transition graph is the graph having $\{0,1\}^n$ as vertex set and the edges set $\{x \rightarrow \bar{x}^{\{u\}} \mid x \in \{0,1\}^n, u \in [n], x_u \neq (f(x))_u\}$. An existing path from x to y is noted $x \rightarrow^* y$.*

Figure 2 gives the transition graph of the asynchronous dynamics of Boolean network of Example 1.

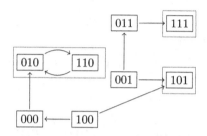

Fig. 2. Transition graph of the Boolean network defined in Example 1. We use shorter notations, 010 meaning that the node 1 has 0 as value, the node 2 has 1 as value, and the node 3 has 0 as value. The attractors are boxed in magenta

Attractors, Fixed point: BN's *Attractors* are the terminal strongly connected components of the transition graph, and can be seen as the long-term dynamics of the system. Note that an attractor is always a set of states, but it can contain either multiple distinct nodes, that is the system oscillate between multiple states (*cyclic attractor*) or a unique point, i.e. the system stays in the same state (*fixed point*).

Definition 5 (Attractor)

$$S \subseteq \{0,1\}^n \text{ is an attractor } \Leftrightarrow S \neq \emptyset \qquad (1)$$

$$and \; \forall x \in S, \forall y \in \{0,1\}^n \setminus S, x \nrightarrow y \qquad (2)$$

$$and \; \forall x \in S, S \setminus x \text{ does not verify (2)} \qquad (3)$$

If $|S| = 1$ then S is a fixed point. Otherwise S is a cyclic attractor.

Given a BN f, $\mathrm{FP}(f) \subseteq \{0,1\}^n$ denotes the set of its fixed points ($\forall x \in \mathrm{FP}(f), f(x) = x$).

Example 2. The BN of Example 1 has 3 attractors that correspond to the 3 terminal strongly connected components of Fig. 2: $\{010, 110\}$ (cyclic attractor), $\{101\}$ and $\{111\}$ (fixed points).

2.2 On the Link Between Attractors and the Interaction Graph

Theorem 1 is a conjecture by René Thomas [16] that has been since demonstrated for Boolean and discrete networks [2,17]: if a Boolean network has multiple attractors then its interaction graph necessarily contains a positive cycle. In the case of multiple fixed points, any pair of fixed point differ at least on a set of nodes forming a positive cycle.

Theorem 1 (Thomas' first rule). *If $G = (V, E)$ has no positive cycles, then f has at most one attractor. Moreover, if f has two distinct fixed points x and y, then G has a positive cycle between vertices $C \subseteq V$ such that $x_v \neq y_v$ for every vertex v in C.*

We can also remark that for a vertex to stay at a value y_v where y is a fixed point, it only needs its ancestors to have the same values as in y.

Remark 1. $\forall y \in \mathrm{FP}(f), \forall u \in [n], \forall z \in \{0,1\}^n$, z verifying $\forall j \in P_u$, $z_j = y_j$, we have $f_u(y) = y_u = f_u(z)$.

Proof. Let u be a vertex in $[n]$. $f(u)$ only depends of the incoming arcs in u, so it only depends of p_u, which in turn depends on its parents. By induction, $f_u(y)$ only depends of P_u, and so, if $f_u(y) = y_u$ in G, then $f_u(y) = y_u$ in $G[P_u]$.   □

3   Formalisation of the BN Reprogramming with Permanent Perturbations

Given two fixed points x and y of Boolean network f, we want to identify sets of nodes, referred to as Reprogramming Determinants (RDs), that when changed in x enable to switch to y. As our theorems rely on the differences between the fixed points, we chose to focus on fixed points solely. Further work will extend, if possible, these theorems and algorithms to all kind of attractors. In the scope of this paper, by "change" we mean permanently set the vertex to a new fixed value. If we "change" u to 1 (resp. 0), then $f_u(x) = 1$ (resp. 0) for all x. When switching to y (by changing I) is possible, we have two cases: it either means that y is reachable from $x_{[x_I = y_I]}$ (existential reachability, Definition 6), or that y is the only reachable fixed point from $x_{[x_I = y_I]}$ (inevitable reachability, Definition 7). These are two different approaches that we will both consider. To remove the temporal aspect, we make all the changes at the same time (hence $x_{[x_I = y_I]}$, otherwise an order should be visible), and only watch if y is reachable. This also means that there is no indication of how long it takes for y to be reached.

Definition 6 (Existential Reachability). *With the boolean network F, a function ER_F can be defined as $ER_F : 2^{2^{[n]}}$, with $ER_F(x, y) \mapsto v$ where v is the set of all minimal vertex sets I such that $x_{[x_I = y_I]} \to^* y$.*

Definition 7 (Inevitable Reachability). *Similarly, a function $IR_F : 2^{2^{[n]}}$ can be defined as $IR_F(x, y) \mapsto w$ where w is the set of all minimal vertices sets I such as $\forall z \in \{0,1\}^n$, $x_{[x_I = y_I]} \to^* z \Rightarrow z \to^* y$.*

These two functions will give different results, and have different meanings, as shown in the example below.

Example 3. Let us consider the BN f of Fig. 3 and its transition graph reproduced in Fig. 4. f has 4 fixed points: $0000, 0001, 1100$ and 1101. Let $x = 0000$ and $y = 1100$. Fixing the node $\{1\}$ to 1 in x makes y reachable: 1100 $(=y)$ is reachable from $x_{[x_1=1]} = 1000$ with the Boolean network f' defined by $f'_1(x) = 1$ and $f'_2 = f_2$, $f'_3 = f_3$, $f'_4 = f_4$. The transition graph of f', considering the first node being active, corresponds to the left part of the transition graph in Fig. 4. One can then remark that y is not the only fixed point reachable: from 1000, 1101 is also reachable. If we also fix the node $\{4\}$ to 0, y is the only reachable fixed point from $x_{[x_1=1,x_4=0]}$ in the Boolean network f'' such that $f''_1(x) = 1$, $f''_2 = f_2$, $f''_3 = f_3$, and $f''_4(x) = 0$.

Therefore, with the previous definitions, $\{1\} \in ER_F(0000, 1100)$ but $\{1\} \notin IR_F(0000, 1100)$; and $\{1, 4\} \in IR_F(0000, 1100)$ but $\{1, 4\} \notin ER_F(0000, 1100)$. Moreover, we also have $\{1, 2\}$ and $\{1, 3\} \in IR_F(0000, 1100)$.

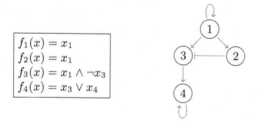

Fig. 3. A BN of dimension 4

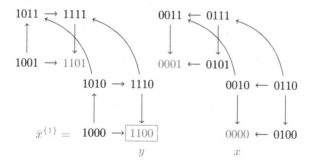

Fig. 4. Transition graph of the BN in Fig. 3

4   Reprogramming Determinants and the SCCs of the Interaction Graph

In this section, we show the link between the RDs and the Strongly Connected Components (SCCs) of the interaction graph of the Boolean network f. Our results make the assumption that all the attractors of f are fixed points (no cyclic attractors).

4.1   SCC Ordering

To switch from x to y, we want to change the value of each vertex u that has different values for x and y ($x_u \neq y_u$) and to prevent each vertex v that verifies $x_v = y_v$ from changing value. We know that changing the value of a vertex can have an impact on other vertices, but we also know that it will only impact its descendants.

So, if a vertex has a different value in x and y but none of its ancestors do, then it is necessary to change this vertex. So, to know which vertices need to be changed first, the best way is to order them, with a topological order for example. Of course, if there are loops, an order is impossible to determine, we have to reduce all SCCs to single "super-vertices" to achieve it. In the remaining of this paper, we will consider SCCs which contain at least one positive cycle, because they are known to change between fixed points (Theorem 1), we call $\mathcal{O}$ the SCC set that contains all such SCCs. Reducing the graph to its SCCs makes possible to rank them from 1 to k with any topological order, noted $\prec$: for all $i, j \in [k], j > i \Rightarrow \mathcal{O}_j \nprec \mathcal{O}_i$.

Let C_0 be the set $\{\mathcal{O}_i \in \mathcal{O} \mid \nexists \mathcal{O}_j, \mathcal{O}_j \prec \mathcal{O}_i\}$, and recursively define slices $C_K = \{\mathcal{O}_i \in (\mathcal{O} \setminus \bigcup_{l \in \{1,..,K-1\}} C_l) \mid \nexists \mathcal{O}_j, \mathcal{O}_j \prec \mathcal{O}_i\}$. Given the definition of the slices, for all topological orders, the slice set will be the same. The slices are numbered from 1 to c.

From this order, we know which SCCs need to be impacted, still, SCCs ranked lower in the hierarchy need not be impacted by the change in their ancestors (see Example 4) The relation $\prec$ only gives an order to make the changes, from which one can determine if further changes are needed.

Example 4. Showing that only using the topological order is not sufficient.

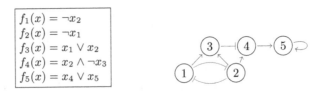

$$
\begin{array}{l}
f_1(x) = \neg x_2 \\
f_2(x) = \neg x_1 \\
f_3(x) = x_1 \vee x_2 \\
f_4(x) = x_2 \wedge \neg x_3 \\
f_5(x) = x_4 \vee x_5
\end{array}
$$

Fig. 5. BN preventing changes in the lower SCC

Any algorithm that only used the topological order without computing the reachable fixed points would not suffice, as the example from Fig. 5 shows: the switch from the fixed point 01100 to 10101 would be computed by just modifying $\{1\}$, but in fact $\{4\}$ will always be fixed at 0, because $\{4\}$ is always inhibited by $\{3\}$, so $\{5\}$ needs to be changed too.

4.2   SCC Filtering

Whether we want y to be the only reachable attractor, or merely to be one of potential several such attractors, the ordering from the previous part is the same, but the filtering will differ.

Theorem 2. *If a vertex u such as $x_u \neq y_u$ and u is not in a positive cycle, then modifying u's ancestors is sufficient to modify u.*

More generally, to switch from x to y, modifying only those strongly connected components that contain at least a positive cycle is sufficient.

Proof. Let u be a vertex such that $x_u \neq y_u$ and u does not lie in a positive cycle. If u is in a negative cycle, the incoming arc from the cycle is irrelevant: given that x and y are fixed points and that u has a distinct value in each, the negative cycle does not change u's value. Given that u is not in a positive cycle, u is not in a SCC (or not relevant if it is in a negative cycle). That means that none of the ancestors are descendants of u. Let z be the state where all of P_u (u's ancestors) have the same value that in y. By the remark from Sect. 2, for all $v \in G[P_u]$, we have $f_v(z) = z_v = y_v$. So, either $f_u(z) = y_u$, and the theorem is proven, either $f_u(z) \neq y_u$, then, by Theorem 1, u is in a positive cycle, contradiction.   □

By recursion over the first part, modifying all the SCCs that contain positive cycles so their vertices have the same value as in y modifies all their children, and then all the children of their children, and so on, until the whole graph has the same values as y.   □

Selecting the SCCs will differ with the two methods. It relies on the same base, searching the higher SCC that should have its values modified and that is not already selected. "Modified" means that all the values of the SCC are fixed to their values in y. The set of the selected SCCs is $\mathcal{S}$.

4.3   SCC Filtering for Existential Reachability

We consider the RDs for the BN reprogramming with Existential Reachability. We give an algorithm to identify different sets of SCCs for which the mutation in the initial fixed point ensure the reachability of the target fixed point. We will prove that the identified combination of SCCs is complete and minimal.

Basically, the algorithm reviews linearly the SCC slices according to $\prec$ and adds the minimal combinations of SCCs to $\mathcal{S}$ that are different in y and the fixed points reachable from $x_{[x_S = y_S]}$:

1. $S := \emptyset$
2. For i ranging from 1 to c:
   - $T := \emptyset$
   - $\forall s \in P(C_i)$ such that s minimal
     $\exists z \in \{0,1\}^n$, $z_{C_i \setminus s} = y_{C_i \setminus s}$, $x_{[x_I = y_I | I \in s]} \to^* z$, $T := T \cup s$.
   - $S := S \bar{\times} T$.

With $\bar{\times}$ being a product and union: for a set I of subsets $I_1, .., I_k$ and a set $J_1, .., J_l$, this product $\bar{\times}$ is defined by: $I \bar{\times} J = \{I_1 \cup J_1, .., I_1 \cup J_l, I_2 \cup J_1, ..., I_k \cup J_l\}$

Complexity: In the worst case, the above algorithm perform $c \times 2^l$ reachability checks (PSPACE-complete [5]), where l is the size of the largest slice.

Existence of a solution and proof of correctness: Forcing all SCCs of such problem that differ on x and y to have the same value as in y is one solution. In the worst case, that is what the algorithm will find. Since the algorithm tests reachability, and a solution exists, it will find one.

Example 5. We apply the algorithm on the BN of Fig. 6 with $x = 00000$ and $y = 11011$.

1. $S := \emptyset$
2. C_1: s minimal $\Leftrightarrow s = \{1\}$
3. $S := S \bar{\times} \{1\} = \{\{1\}\}$
4. C_2: s minimal $\Leftrightarrow s = \emptyset$ [1]
5. $S := S \bar{\times} \emptyset = \{\{1\}\}$.

We now prove the completeness of the algorithm and the minimality of the returned sets of SCCs (Theorem 3) and that any RDs in $ER(x, y)$ is spans only and necessarily in one of the set of SCCs identified by the algorithm (Theorem 4).

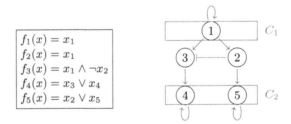

$$
\begin{array}{l}
f_1(x) = x_1 \\
f_2(x) = x_1 \\
f_3(x) = x_1 \wedge \neg x_2 \\
f_4(x) = x_3 \vee x_4 \\
f_5(x) = x_2 \vee x_5
\end{array}
$$

Fig. 6. BN of dimension 5 (left) with its interaction graph (right). Slices are enclosed in boxes. $C_1 = \{\{1\}\}$, $C_2 = \{\{4\}, \{5\}\}$

Theorem 3. *S only contains minimal SCC sets, and S is complete.*

Proof. **Minimality:** Inside every slice, the SCCs are totally independent one another. Moreover, given the order exploiting, we can deduce that the sum of the minima on each slice is the minimum on the whole graph.   □

---

[1] with the path $10000 \to 10100 \to 10110 \to 11110 \to 11111$ (and the fixed point is the next step, $\to 11011$ but there is no need to go further than 11111.).

Completeness: Let I be a minimal SCC set such as $x_{[x_J=y_J|J\in I]} \to^* y$, then, for every slice C_i, $I \cap C_i$ is minimal, since once all the SCCs in a slice can be changed to the way they are in y, we can always choose the path that allows this change. Hence $I \in \mathcal{S}$. ☐

Theorem 4. $\forall c \in ER(x,y), \exists I \in \mathcal{S}, \forall u \in c, \exists scc \in I, u \in scc.$

Proof. Let c be a vertex set in $ER(x,y)$ and u one of the vertices. If $u \notin \mathcal{O}$, then c is not minimal, by Theorem 2. If for all $I \in \mathcal{S}$, u is in $o \in (\mathcal{O} \setminus I)$ then there exists a path such that changing o's ancestors makes o's change possible, and the ancestors need to be changed as well, by construction of I. So $c \setminus u$ would have the same effect, and c would not be minimal. If $u \notin o$, then there exists $I \in \mathcal{S}$ and $scc \in I$, such as $u \in scc$. ☐

4.4   SCC Filtering for Inevitable Reachability

We now give an algorithm to identify a set of SCCs for which the mutation in the initial fixed point is sufficient to ensure the *Inevitable Reachability* of the target fixed point.

The algorithm computes all reachable fixed points from x with the SCCs in $\mathcal{S}$ modified, and find the one, z, that has the lower SCC (in the ranking given by $\prec$) in which a vertex u is such that $z_u \neq y_u$. As we are looking for all reachable fixed points, this will always return the same SCC (even if the order is only partial), thus allowing the algorithm to be deterministic. We add this SCC to $\mathcal{S}$, and repeat until y is the only reachable fixed point.

1. $\mathcal{S} := \emptyset$
2. While $\exists z \in \mathrm{FP}(f)$, $z \neq y$, $x_{[x_I=y_I|I\in\mathcal{S}]} \to^* z$
   - $\mathcal{S} := \mathcal{S} \cup \{\mathcal{O}_i\}$, with
     $i = min_{a\in\{1,..,k\}}(a \mid \exists z \in \mathrm{FP}(f), z_{\mathcal{O}_a} \neq y_{\mathcal{O}_a}, x_{[x_I=y_I|I\in\mathcal{S}]} \to^* z)$

If two (or more) SCCs A and B are such that they are differently ordered in two distinct orders, then A has no influence on B and neither has B on A. Then, the algorithm will select both SCCs if neither are impacted by the previous changes, so the order does not matter.

Existence of a solution and proof of correctness: A solution is to fix all the SCCs of the graph to their value in y. Since there exists a solution and the algorithm tests if y is the only reachable point, and follows the order given by $\prec$, it will end and find a solution.

Complexity: Computing all fixed points reachable is PSPACE-complete [4]. It is used k times (number of SCCs) in the worst case.

Example 6. We apply the algorithm on the BN of Fig. 7. with $x = 00000$ and $y = 11011$. Starting from $\mathcal{S} := \emptyset$, the only reachable fixed point is $(0)00(0)(0)$ (the SCCs from $\mathcal{O}$ are parenthesized). The smallest SCC o such as $x_{[x_S=y_S],o} \neq y_o$ is $\mathcal{O}_1$, so $\mathcal{S} := \emptyset \cup \{\mathcal{O}_1\} = \{\mathcal{O}_1\}$. The reachable fixed points from

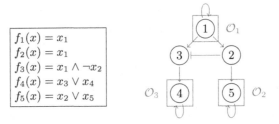

Fig. 7. BN of dimension 5 (left) with its interaction graph (right) on which the SCCs containing positive cycles ($\mathcal{O}$) are boxed

$x_{[x_I=y_I|I\in\mathcal{S}]} = 10000$ are now: $(1)10(1)(1)$ and $(1)10(0)(1)$. The smallest SCC o such that $x_{[x_\mathcal{S}=y_\mathcal{S}],o} \neq y_o$ is $\mathcal{O}_3$. We set $\mathcal{S} := \mathcal{S} \cup \mathcal{O}_3 = \{\mathcal{O}_1, \mathcal{O}_3\}$ and we obtain that the only reachable fixed point from $x_{[x_I=y_I|I\in\mathcal{S}]} = 10010$ is $(1)10(1)(1)$ which is y. So the algorithm stops.

Theorem 5. $\mathcal{S}$ *is minimal.*

Proof. If a set S_1 exists such that S_1 has a lower cardinal than $\mathcal{S}$ and modifying S_1 makes y the only reachable point, then we can reduce S_1 to a subset of $\mathcal{S}$. Let s be a SCC in $\mathcal{S} \setminus S_1$, thus there exists a fixed point z such that $z_s \neq y_s$ and by construction of $\mathcal{S}$, z is reachable from x modified by S_1.   □

We remark that, contrary to the case of Existential Reachability, the RDs for Inevitable Reachability of the target fixed point are not necessarily in SCCs containing positive cycles. Indeed, in Example 3, we showed that $IR_F(x,y)$ can refer to nodes that do not belong to $\mathcal{O}$ (such as the node 2 for the BN of Fig. 3). But we can also remark that if a RD v is not in a SCC containing a positive cycle, then $x_v = y_v$.

Theorem 6. $\forall v \in IR(x,y), x_v \neq y_v \Rightarrow \exists scc \in \mathcal{O}, v \in scc.$

Proof. Let $v \in IR(x,y)$ such that for all $scc \in \mathcal{O}$, $v \notin scc$. By Theorem 1, if v is such that $x_v \neq y_v$, then modifying the SCC in $\mathcal{O}$ is enough to modify v. But $v \in IR(x,y)$ and $IR(x,y)$ is minimal, thus $x_v = y_v$.   □

5   Identifying Determinants Within SCCs

We know that modifying all the SCCs selected is enough to switch from x to y, but to reduce the genes selected, we could try to modify only some of the vertices to achieve the same result. But, as dynamics are involved, there could be unwanted changes (or wanted and unpredicted changes, in the case where we want y to be reachable) in the descendants.

An idea could be to select the feedback vertex set of the SCC: by fixing the vertices from this set, we effectively destroy every circle, thus the only reachable state of the SCC is the one having the same values as y. This, however, does

$$
\begin{aligned}
f_1(x) &= \neg x_3 \wedge \neg x_2 \\
f_2(x) &= \neg x_1 \\
f_3(x) &= \neg x_1 \\
f_4(x) &= x_2 \wedge \neg x_1 \wedge \neg x_3 \\
f_5(x) &= x_4 \vee x_5
\end{aligned}
$$

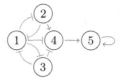

Fig. 8. BN of dimension 5 (left) and its interaction graph (right)

not solve the problem: in Example 7, $\{1\}$ is the feedback vertex set, and we still have the same issue. Moreover, it miss some of the possible solutions (modifying $\{2\}$ or $\{3\}$ could work to change the whole SCC in Example 7) or even dismiss the best solution (in Example 7, changing $\{3\}$ makes y the only reachable fixed point and solves the issue).

Example 7. Illustration of the problem with dynamics.

We decide that $x = 10000$ and $y = 01100$, and $01101 = z$, those are all fixed points. Let's suppose we want y to be the only fixed point reachable. The algorithm will see that if the whole first SCC is modified $\{1, 2, 3\}$, y is the only reachable fixed point. It could pick $\{1\}$ to be modified, but instead of $00000 \to 00100 \to 01100$, we can have

$$00000 \to 01000 \to 01010 \to 01011 \to 01111 \to 01101$$

This leads to z being reachable by only modifying $\{1\}$, and so the algorithm would be wrong.

This leaves to two kinds of approaches: either a way to modify the SCC so that it does not impact its descendants can be found, either we need to select the vertices to be modified in the SCC as a intermediary step in the process, and redesign $\mathcal{S}$ as a list of vertices instead of a list of SCCs.

By exploiting the results of the preceding section, we show an algorithm to compute a set of RDs which guarantees the Inevitable Reachability of the target fixed point. The algorithm recursively picks a vertex u in the lowest SCC in the order given by $\prec$ in $\mathcal{O}$, and modify its associated function to become the constant value y_u. The interaction graph of the resulting Boolean network is a sub-graph of the initial interaction graph, where all the input edges of the node u have been removed. Hence, the SCC $\mathcal{O}_1$ is split in the new interaction graph. If necessary, another vertex can be picked in the lowest SCC in the new interaction graph:

RecursiveAlgorithm(f, rd):

- If $\exists z \in \mathrm{FP}(f)$, $x_{[x_{rd}=y_{rd}]} \to^* z$ then:
  - $res = \emptyset$
  - $i = \min_{a \in \{1,..,k\}} (a \mid \exists z \in \mathrm{FP}(f), z_{\mathcal{O}_a} \neq y_{\mathcal{O}_a}, x_{[x_I=y_I \mid I \in \mathcal{S}]} \to^* z)$

- For all $u \in \mathcal{O}_i$:
    - $* \; g := f$ with $g_u := y_u$
    - $* \; res := res \cup$ RecursiveAlgorithm$(g, rd \bar{\times} \{u\})$
- return res
- else:
  - return rd

Remark that the algorithm always find at least one solution: if the target fixed point is not the only reachable fixed point, then there is at least one positive cycle (and hence a SCC) which has a different state (and hence will be selected by our algorithm).

Example 8. Applied to the BN of Fig. 8 with $x = 10000$ and $y = 01100$, the above algorithm returns, for instance, the RD $\{2, 5\}$: indeed, $\{2\}$ belongs to $\mathcal{O}_1$. When fixing $f_2 = 1$, the new interaction graph has two SCCs with positive cycles: $\{1, 3\}$ and $\{5\}$. From the state 11000, two fixed points are reachable: 01100, 01101. Hence, because the SCC $\{1, 3\}$ has the same values than in y in those two fixed points, the next vertex in picked in the SCC $\{5\}$. Finally, from the state 11001, y is the only reachable fixed point.

6   Discussion

This paper provides the first formal characterization of the Reprogramming Determinants (RDs) for switching from one fixed point to another in the scope of the asynchronous dynamics of Boolean networks.

In the case of reprogramming with existential reachability of the target fixed point, we prove that all the possible minimal RDs modify nodes in particular combinations of SCCs of the interaction graph of the Boolean network. We give an algorithm to determine exactly those combinations of set of nodes. Our characterizations rely on the verification of reachability properties.

In the case of reprogramming with inevitable reachability of the target fixed point, we show that the RDs are not necessarily in SCCs. However, we provide an algorithm which identifies RDs that guarantee the inevitable reachability by picking nodes in appropriate SCCs. The algorithm relies on the enumeration of reachable fixed points.

One of the main limitation of our algorithms is the numerous reachability checks it needs to perform. Future work will consider methods and data structures for factorizing the exploration of the Boolean network dynamics.

The present work considered only permanent mutations: when a node is mutated, it is assumed it keeps its mutated value forever (its local Boolean function becomes a constant function). Considering temporary mutations, i.e., where the local Boolean function of mutated nodes is restored after some time, is a challenging research direction: one should determine the ordering and the duration of mutations, and the set of candidate mutations is *a priori* no longer restricted to connected components, as it is the case for permanent mutations.

References

1. Abou-Jaoudé, W., Monteiro, P.T., Naldi, A., Grandclaudon, M., Soumelis, V., Chaouiya, C., Thieffry, D.: Model checking to assess T-helper cell plasticity. Front. Bioeng. Biotechnol. **2** (2015)
2. Aracena, J.: Maximum number of fixed points in regulatory boolean networks. Bull. Math. Biol. **70**(5), 1398–1409 (2008)
3. Chang, R., Shoemaker, R., Wang, W.: Systematic search for recipes to generate induced pluripotent stem cells. PLoS Comput. Biol. **7**(12), e1002300 (2011)
4. Chatain, T., Haar, S., Jezequel, L., Paulevé, L., Schwoon, S.: Characterization of reachable attractors using petri net unfoldings. In: Mendes, P., Dada, J.O., Smallbone, K. (eds.) CMSB 2014. LNCS, vol. 8859, pp. 129–142. Springer, Heidelberg (2014). doi:10.1007/978-3-319-12982-2_10
5. Cheng, A., Esparza, J., Palsberg, J.: Complexity results for 1-safe nets. Theor. Comput. Sci. **147**(1&2), 117–136 (1995)
6. Crespo, I., Perumal, T.M., Jurkowski, W., del Sol, A.: Detecting cellular reprogramming determinants by differential stability analysis of gene regulatory networks. BMC Syst. Biol. **7**(1), 140 (2013)
7. del Sol, A., Buckley, N.J.: Concise review: a population shift view of cellular reprogramming. Stem Cells **32**(6), 1367–1372 (2014)
8. Gao, Z., Chen, X., Başar, T.: On the stability of conjunctive boolean networks, March 2016
9. Graf, T., Enver, T.: Forcing cells to change lineages. Nature **462**(7273), 587–594 (2009)
10. Jo, J., Hwang, S., Kim, H.J., Hong, S., Lee, J.E., Lee, S.-G., Baek, A., Han, H., Lee, J.I., Lee, I., et al.: An integrated systems biology approach identifies positive cofactor 4 as a factor that increases reprogramming efficiency. Nucleic Acids Res. **44**(3), 1203–1215 (2016)
11. Paulevé, L.: Goal-oriented reduction of automata networks. In: Bartocci, E., Lio, P., Paoletti, N., Ogbuji, B. (eds.) CMSB 2016. LNCS, vol. 9859, pp. 252–272. Springer, Heidelberg (2016). doi:10.1007/978-3-319-45177-0_16
12. Rackham, O.J.L., Firas, J., Fang, H., Oates, M.E., Holmes, M.L., Knaupp, A.S., Suzuki, H., Nefzger, C.M., Daub, C.O., Shin, J.W., Petretto, E., Forrest, A.R.R., Hayashizaki, Y., Polo, J.M., Gough, J.: A predictive computational framework for direct reprogramming between human cell types. Nat. Genet. **48**(3), 331–335 (2016)
13. Richard, A.: Positive circuits and maximal number of fixed points in discrete dynamical systems. Discrete Appl. Math. **157**(15), 3281–3288 (2009)
14. Richard, A.: Negative circuits and sustained oscillations in asynchronous automata networks. Adv. Appl. Math. **44**(4), 378–392 (2010)
15. Takahashi, K., Yamanaka, S.: A decade of transcription factor-mediated reprogramming to pluripotency. Nat. Rev. Mol. Cell. Biol. **17**(3), 183–193 (2016)
16. Thomas, R.: Boolean formalization of genetic control circuits. J. Theor. Biol. **42**(3), 563–585 (1973)
17. Remy, É., Ruet, P., Thieffry, D.: Graphic requirements for multistability and attractive cycles in a boolean dynamical framework. Adv. Appl. Math. **41**(3), 335–350 (2008)

Discrete Abstraction of Multiaffine Systems

Hui Kong[1([⊠])], Ezio Bartocci[2], Sergiy Bogomolov[1], Radu Grosu[2],
Thomas A. Henzinger[1], Yu Jiang[3], and Christian Schilling[4]

[1] Institute of Science and Technology Austria, Klosterneuburg, Austria
hui.kong@ist.ac.at
[2] Vienna University of Technology, Vienna, Austria
[3] University of Illinois at Urbana-Champaign, Champaign, IL, USA
[4] University of Freiburg, Freiburg im Breisgau, Germany

Abstract. Many biological systems can be modeled as multiaffine
hybrid systems. Due to the nonlinearity of multiaffine systems, it is difficult to verify their properties of interest directly. A common strategy to
tackle this problem is to construct and analyze a discrete overapproximation of the original system. However, the conservativeness of a discrete
abstraction significantly determines the level of confidence we can have
in the properties of the original system. In this paper, in order to reduce
the conservativeness of a discrete abstraction, we propose a new method
based on a sufficient and necessary decision condition for computing discrete transitions between states in the abstract system. We assume the
state space partition of a multiaffine system to be based on a set of multivariate polynomials. Hence, a rectangular partition defined in terms of
polynomials of the form $(x_i - c)$ is just a simple case of multivariate
polynomial partition, and the new decision condition applies naturally.
We analyze and demonstrate the improvement of our method over the
existing methods using some examples.

Keywords: Multiaffine system · Hybrid system · Discrete abstraction ·
State space partition · Gröbner basis

1 Introduction

A biological system is a complex network of biologically relevant entities. The
analysis of complex biological systems can significantly benefit from the theory
and techniques developed in the area of hybrid systems [4–7,11,12,15,16,21,22].
The class of multiaffine hybrid systems [9,17] is particularly suited to model
and analyze a broad range of biological systems. However, due to the nonlinearity of multiaffine systems, it is often difficult to verify their properties of
interest directly. A common strategy to tackle this problem is based on the idea
of hybridization. In this setting, a given system is replaced by an abstraction
where the system state space is partitioned and the original nonlinear dynamics
is replaced with a simpler one in each induced partition. The resulting abstraction can either keep some approximated version of continuous dynamics [2,3]

© Springer International Publishing AG 2016
E. Cinquemani and A. Donzé (Eds.): HSB 2016, LNBI 9957, pp. 128–144, 2016.
DOI: 10.1007/978-3-319-47151-8_9

or reason in discrete terms only [18–20]. In the following, we consider discrete abstractions of hybrid systems.

The quality of a discrete abstraction of a multiaffine system depends closely on the partition scheme of the state space and the conservativeness of discrete transitions between abstract states. A simple idea to partition the state space is to use a set of hyperplanes that are perpendicular to coordinate axes [8,13,23], hence the resulting regions are a set of hyperrectangles. The benefits of rectangular partition can be described as follows: (1) vertices of the hyperrectangles can be easily obtained, (2) some properties can be applied to establish the discrete transitions between abstract states (e.g., Proposition 1). However, since a rectangular partition does not take into account the feature of the vector flow, it could be inefficient. To address this problem, in [1,28–30], a set of polynomials was used for partitioning the continuous state space. The idea is that, given a set of polynomials $\Phi = \{\varphi_i(\boldsymbol{x}) \in \mathbb{R}[\boldsymbol{x}], i = 1, \ldots, K\}$, each $\varphi_i(\boldsymbol{x})$ can partition the state space into three parts: (1) $\{\boldsymbol{x} \in \mathbb{R}^n \mid \varphi_i(\boldsymbol{x}) < 0\}$, (2) $\{\boldsymbol{x} \in \mathbb{R}^n \mid \varphi_i(\boldsymbol{x}) = 0\}$ and (3) $\{\boldsymbol{x} \in \mathbb{R}^n \mid \varphi_i(\boldsymbol{x}) > 0\}$. Thus, $|\Phi|$ polynomials altogether can partition the state space into at most $3^{|\Phi|}$ parts. Both of the aforementioned partition methods have to address an important issue: how to establish discrete transitions between the abstract states (i.e. partitioned regions)? A common decision condition used by the existing methods is that, a positive first-order Lie derivative of $\varphi_i(\boldsymbol{x})$ at some point $\boldsymbol{x}_\tau$ in the hypersurface $\varphi_i(\boldsymbol{x}) = 0$ suffices to prove a trajectory being able to reach the region of $\varphi_i(\boldsymbol{x}) > 0$ from the region of $\varphi_i(\boldsymbol{x}) < 0$ and vice versa. However, this conditional test sometimes fails (i.e. the first-order Lie derivative is 0) and an overapproximating transition relation has to be built.

In this paper, similar to [28,29], the state space partition is assumed to be based on a set Φ of multivariate polynomials. To reduce the conservativeness of a discrete abstraction of multiaffine system, we propose a necessary and sufficient condition to build the discrete transitions between the abstract states. The idea is that, given a hypersurface $H_{\varphi_i} = \{\boldsymbol{x} \in \mathbb{R}^n \mid \varphi_i(\boldsymbol{x}) = 0\}$ with $\varphi_i \in \Phi$, a trajectory can pass through H_{φ_i} at some $x_\tau \in H_{\varphi_i}$ if and only if there exists an odd number N such that the N'th-order Lie derivative $\mathcal{L}_f^N \varphi_i$ of φ_i is not equal to 0 and all the i'th-order Lie derivative (for $1 < i < N$) of φ_i is 0 at x_τ, and fortunately, there is an upper bound for N which is computable using the Gröbner basis. More specifically, the direction of the trajectory relative to H_{φ_i} at x_τ depends on the sign of $\mathcal{L}_f^N \varphi_i$: if $\mathcal{L}_f^N \varphi_i > 0$, the trajectory moves from the region of $\varphi_i(\boldsymbol{x}) < 0$ to the region of $\varphi_i(\boldsymbol{x}) > 0$, otherwise, the direction reverses. For any two adjacent abstract states $\boldsymbol{u}$ and $\boldsymbol{v}$ (see Definition 4 for *adjacency*), the problem of deciding the transition relation between them is equivalent to deciding whether there exists a trajectory that passes through the intersection of multiple hypersurfaces, which can be formalized as a first-order logic formula consisting of Lie derivatives of $\varphi_i(\boldsymbol{x})$ and can be solved by an SMT solver.

The main contribution of this paper includes: (1) we propose a necessary and sufficient condition for building discrete transitions between abstract states, (2) we design an algorithm for establishing the transition relations between abstract states, (3) we analyze and demonstrate the improvement of our method over the existing methods.

The rest of the paper is organized as follows. Section 2 gives the preliminaries required for the paper. Section 3 describes the partition scheme and the mapping between the abstract states and the original state regions. Section 4 proposes the method for establishing discrete transitions between abstract states. In Sect. 5, we analyze and demonstrate the improvement of our method over existing methods. Finally, we conclude in Sect. 6.

2   Preliminaries

In this section, we recall some backgrounds we need throughout the paper. We first clarify some notation conventions. If not specified otherwise, we decorate vectors in bold face (e.g., $\boldsymbol{x}$), we use the symbol $\mathbb{K}$ for a field, $\mathbb{R}$ for the real number field, $\mathbb{C}$ for the complex number field (which is algebraically closed) and $\mathbb{N}$ for the set of natural numbers, and all the polynomials involved are multivariate polynomials. In addition, for all the polynomials $p(\boldsymbol{x})$, we denote by $\boldsymbol{x}$ the vector composed of all the variables that occur in the polynomial. $|\Psi|$ denotes the cardinality of the set Ψ.

Definition 1 [14]. *A subset $I \subseteq \mathbb{K}[\boldsymbol{x}]$ is called an ideal if*

1. *$0 \in I$,*
2. *if $p(\boldsymbol{x}), q(\boldsymbol{x}) \in I$, then $p(\boldsymbol{x}) + q(\boldsymbol{x}) \in I$,*
3. *if $p(\boldsymbol{x}) \in I$ and $g(\boldsymbol{x}) \in \mathbb{K}[\boldsymbol{x}]$, then $p(\boldsymbol{x})g(\boldsymbol{x}) \in I$.*

Definition 2 [14]. *Let $g_1, ..., g_s$ be polynomials in $\mathbb{K}[\boldsymbol{x}]$, where $\mathbb{K}$ is a field. Then we set*

$$\langle g_1, ..., g_s \rangle = \left\{ \sum_{i=1}^{s} h_i g_i : h_1, ..., h_s \in \mathbb{K}[\boldsymbol{x}] \right\} \tag{1}$$

It is easy to verify that $\langle g_1, ..., g_s \rangle$ is an ideal and it is called the ideal generated by $\{g_1, ..., g_s\}$.

For the denotative convenience, we need to first present the notation of Lie derivative, which is widely used in the discipline of differential geometry. For a given polynomial $\varphi \in \mathbb{K}[\boldsymbol{x}]$ and a continuous system $\dot{\boldsymbol{x}} = \boldsymbol{f}$ (where $\boldsymbol{f} = (f_1, ..., f_n)$), the high-order Lie derivative of φ is defined as follows.

$$\mathcal{L}_f^k \varphi \triangleq \begin{cases} \varphi, & k = 0 \\ \sum_{i=1}^{n} \frac{\partial \mathcal{L}_f^{k-1} \varphi}{\partial x_i} \cdot f_i, & k \geq 1 \end{cases}$$

Essentially, the k'th-order Lie derivative of φ is the k'th derivative of φ w.r.t. time t and hence reflects the change of φ over time t. Note that we just write $\mathcal{L}_f^1 \varphi$ as $\mathcal{L}_f \varphi$.

Theorem 1 [27] *(Fixed Point Theorem). Given a polynomial $\varphi \in \mathbb{K}[\boldsymbol{x}]$, if, for some $M > 0, \mathcal{L}_f^{M+1} \varphi \in \langle \mathcal{L}_f^0 \varphi, ..., \mathcal{L}_f^M \varphi \rangle$, then $\forall k \geq M + 1 : \mathcal{L}_f^k \varphi \in \langle \mathcal{L}_f^0 \varphi, ..., \mathcal{L}_f^M \varphi \rangle$.*

Proposition 1 [10]. *Let* $f : R \to \mathbb{R}^q$ *be a multiaffine function on the* n-*dimensional rectangle* $R \subset \mathbb{R}^n$ *and* $\boldsymbol{x} = (x_1, ..., x_n) \in R$, *suppose* F_i *is the lowest-dimensional face of* R *that contains* $\boldsymbol{x}$. *Then,* $f(\boldsymbol{x})$ *is a convex combination of the values of* f *at the vertices of* F_i.

Definition 3 (Multiaffine System). *A* multiaffine System *is a tuple* $M \stackrel{def}{=} \langle X, \boldsymbol{f}, Init \rangle$, *where*

1. X *is the state space of the system* M,
2. $\boldsymbol{f}$ *is a Lipschitz multiaffine polynomial vector flow function, and*
3. *Init is the initial set described by a semialgebraic set.*

A multiaffine polynomial is a polynomial for which if we fix all the variables but one, the polynomial will become a linear polynomial.

3   State Space Partition and Abstract State Mapping

In this section, we introduce the partition scheme we adopt throughout the paper and the mapping of the original states to the abstract states.

3.1   State Space Partition

We assume to use a set of multivariate polynomials to partition the state space. There are several ways available to derive the set of polynomials [24,29,30]: (1) take the polynomials occurring in the vector flow function, the guards and the property to be verified, (2) compute the Lie derivatives of the existing polynomials iteratively, (3) discover algebraic invariants of the system. The details of these techniques are not covered in this paper.

The idea of polynomial-based partition is as follows. Given a set of polynomials $\Phi = \{\varphi_i(\boldsymbol{x}) \in \mathbb{R}[\boldsymbol{x}], i = 1, \ldots, K\}$, each $\varphi_i(\boldsymbol{x})$ can partition the state space into three parts: (1) $\{\boldsymbol{x} \in \mathbb{R}^n \mid \varphi_i(\boldsymbol{x}) < 0\}$, (2) $\{\boldsymbol{x} \in \mathbb{R}^n \mid \varphi_i(\boldsymbol{x}) = 0\}$ and (3) $\{\boldsymbol{x} \in \mathbb{R}^n \mid \varphi_i(\boldsymbol{x}) > 0\}$. Thus, $|\Phi|$ polynomials altogether can partition the state space into at most $3^{|\Phi|}$ regions, each region of the partition can be represented as $\{x \in \mathbb{R}^n \mid \bigwedge \varphi_i(x) \sim_i 0\}$ with $\sim_i \in \{>, =, <\}$. In the following, we describe how to map these regions to abstract states.

3.2   Abstract States Mapping

Given a multiaffine system $C = \langle X, Init, f \rangle$, a polynomial set $\Phi = \{\varphi_i(\boldsymbol{x}) \in \mathbb{R}[\boldsymbol{x}], i = 1, \ldots, K\}$ can partition a state space into at most 3^K regions and every state $\boldsymbol{x} \in X$ can be mapped to an abstract state in $V_\Phi \in 2^{\{-1,0,1\}^K}$ using the following abstraction function $Abst : X \mapsto V_\Phi$.

$$Abst\,(x) = (v_1(\boldsymbol{x}), ..., v_K(\boldsymbol{x})), \quad v_i(\boldsymbol{x}) \triangleq \begin{cases} 1, & \varphi_i(\boldsymbol{x}) > 0 \\ 0, & \varphi_i(\boldsymbol{x}) = 0 \\ -1, & \varphi_i(\boldsymbol{x}) < 0 \end{cases} \quad i = 1, \ldots, K$$

Conversely, we have the following concretization function $Con : V_\Phi \mapsto 2^X$ that maps an abstract state to a region of the original state space.

$$Con\,(v) = \{x \in X \mid \bigwedge_{i=1}^{K} \pi(v_i)\}, \quad \pi(v_i) \triangleq \begin{cases} \varphi_i(x) > 0, & v_i = 1 \\ \varphi_i(x) = 0, & v_i = 0 \\ \varphi_i(x) < 0, & v_i = -1 \end{cases} \quad i = 1, \dots, K$$

In the abstract state space, there could exist discrete transitions only between the abstract states whose corresponding regions in the original state space are **adjacent**, which we now define formally.

Definition 4 (Adjacency). *Given an abstract state space $V_\Phi \in 2^{\{-1,0,1\}^K}$, two abstract states $u, v \in V_\Phi$ are **adjacent**, denoted by $\mathbf{Adj(u,v)}$, if and only if they satisfy the following formula with $\sim \in \{>, <\}$*

$$\begin{aligned} & \mathbf{dim(u)} \neq \mathbf{dim(v)} \wedge \\ & \quad (\mathbf{dim(u)} \sim \mathbf{dim(v)} \implies \forall i = 1, \dots, K : u_i = v_i \vee |u_i| \sim |v_i|) \end{aligned} \quad (2)$$

*where $\mathbf{dim(w)} = \sum_{i=1}^{K} |w_i|$ is called the **dimension** of an abstract state.*

Essentially, Definition 4 means that one of two adjacent abstract states can be obtained by setting some of the nonzero components of the other state to zero. The definition is reasonable because any trajectory $x(t)$ cannot get from the region of $\varphi_i(x) < 0$ to the region of $\varphi_i(x) > 0$ without crossing the hypersurface $\varphi_i(x) = 0$.

Definition 5 (Discrete Abstraction). *Given a multiaffine system $C = \langle X, f, Init \rangle$ and a polynomial set $\Phi = \{\varphi_i(x) \in \mathbb{R}[x], i = 1, \dots, K\}$, a discrete abstraction of C w.r.t. Φ is the transition system $C_\Phi = \langle V_\Phi, T_\Phi, Init_\Phi \rangle$, where*

- *$V_\Phi \in 2^{\{-1,0,1\}^K}$ is the abstract state space;*
- *$T_\Phi \in 2^{V_\Phi \times V_\Phi}$ is the set of discrete transitions such that $(u, v) \in T_\Phi$ if and only if there exists a trajectory $x(t)$ and $t_1, t_2 \in \mathbb{R}_{\geq 0}$ and $t_2 > t_1$ satisfying: (1) $x(t_1) \in Con\,(u)$, (2) $x(t_2) \in Con\,(v)$, (3) $\forall t \in (t_1, t_2) : x(t) \in Con\,(u) \cup Con\,(v)$;*
- *$Init_\Phi = \{v \in V_\Phi \mid \exists x \in Init : x \in Con\,(v)\}$ is the initial set.*

A discrete abstraction is an overapproximation of the original system. Given a partition, to construct a precise discrete abstraction, the key point is to make the set T_Φ of discrete transitions as small as possible. The technique for this purpose is presented in the following section.

4   Establishment of Discrete Transitions

In this section, we introduce how to establish the discrete transitions between the abstract states.

4.1   A Necessary and Sufficient Condition

Suppose we have a polynomial set $\Phi = \{\varphi_i(\boldsymbol{x}) \in \mathbb{R}[\boldsymbol{x}], i = 1, \ldots, K\}$ for the partition, according to Definition 5, for a given pair of abstract states $\boldsymbol{u}$ and $\boldsymbol{v}$. There is a discrete transition from $\boldsymbol{u}$ to $\boldsymbol{v}$ if and only if $\boldsymbol{u}$ and $\boldsymbol{v}$ are adjacent and there exists a trajectory that reaches $Con\,(\boldsymbol{v})$ from $Con\,(\boldsymbol{u})$. Assume $\boldsymbol{u} = (u_1, \cdots, u_K)$ and $\boldsymbol{v} = (v_1, \cdots, v_K)$ are adjacent and let $D_{u,v} = \{\varphi_k(\boldsymbol{x}) = 0 \mid u_k \neq v_k,\ k = 1, \ldots, K\}$, then the original problem is equivalent to deciding whether there exists a trajectory passing through the intersection of all the hypersurfaces in $D_{u,c}$. By the following proposition, we first address the issue of deciding whether there exists a trajectory passing through a single hypersurface.

Theorem 2. *A continuous system $\dot{\boldsymbol{x}} = \boldsymbol{f}(\boldsymbol{x})$ can pass through a hypersurface $H = \{\boldsymbol{x} \in \mathbb{R}^n \mid \varphi(\boldsymbol{x}) = 0, \varphi(\boldsymbol{x}) \in \mathbb{R}[\boldsymbol{x}]\}$, i.e.*

$$\exists \boldsymbol{x}(t) \in \{\boldsymbol{x}(t) \mid \dot{\boldsymbol{x}} = \boldsymbol{f}(\boldsymbol{x}), \boldsymbol{x}(0) \in I_0\} : \exists \tau > 0, \epsilon > 0 : \varphi(\boldsymbol{x}(\tau)) = 0 \\ \wedge\, (\forall t_1 \in (\tau - \epsilon, \tau) : \forall t_2 \in (\tau, \tau + \epsilon) : \varphi(\boldsymbol{x}(t_1))\varphi(\boldsymbol{x}(t_2)) < 0) \tag{3}$$

iff the formula

$$\exists \boldsymbol{x}_\tau \in H : \exists N = 2k - 1 : \bigwedge_{j=1}^{N-1} \mathcal{L}_{\boldsymbol{f}}^j \varphi = 0 \wedge \mathcal{L}_{\boldsymbol{f}}^N \varphi \neq 0 \tag{4}$$

holds, where I_0 is the set of initial states and $k \in \mathbb{N}$. Moreover, if $\mathcal{L}_{\boldsymbol{f}}^N \varphi > 0$, the direction of the trajectory points from the region of $\varphi(\boldsymbol{x}) < 0$ to the region of $\varphi(\boldsymbol{x}) > 0$, and otherwise, the direction reverses.

Proof. The Taylor expansion of $\varphi(\boldsymbol{x}(t))$ at time τ is as follows.

$$\varphi(\boldsymbol{x}(t)) = \varphi(\boldsymbol{x}(\tau)) + \sum_{k=1}^{\infty} \frac{1}{k!}(\mathcal{L}_{\boldsymbol{f}}^k \varphi)(t - \tau)^k \tag{5}$$

$(4) \Rightarrow (3)$: By applying the condition (4) to the Taylor expansion (5), we can derive that $\varphi(\boldsymbol{x}(t)) = \frac{1}{N!}(\mathcal{L}_{\boldsymbol{f}}^N \varphi)(t - \tau)^N$ for some trajectory $\boldsymbol{x}(t)$ with $\boldsymbol{x}(\tau) = \boldsymbol{x}_\tau$. Since $N = 2k - 1$ is an odd number, there must exist a real $\epsilon > 0$ such that,

1. if $\mathcal{L}_{\boldsymbol{f}}^N \varphi > 0$, then $\forall t_1 \in (\tau - \epsilon, \tau) : \varphi(\boldsymbol{x}(t_1)) < 0$ and $\forall t_2 \in (\tau, \tau + \epsilon) : \varphi(\boldsymbol{x}(t_2)) < 0$, or
2. if $\mathcal{L}_{\boldsymbol{f}}^N \varphi < 0$, then $\forall t_1 \in (\tau - \epsilon, \tau) : \varphi(\boldsymbol{x}(t_1)) > 0$ and $\forall t_2 \in (\tau, \tau + \epsilon) : \varphi(\boldsymbol{x}(t_2)) > 0$.

Therefore, condition (3) holds.

$(3) \Rightarrow (4)$: We show this implication by contradiction. Let $M = \min\{j \geq 1 \mid \mathcal{L}_{\boldsymbol{f}}^j \varphi \neq 0\}$. By applying condition (3) to the Taylor expansion (5), we can derive that there exist a real $\epsilon > 0$ such that

$$\forall t \in (\tau - \epsilon, \tau + \epsilon) : \varphi(\boldsymbol{x}(t)) = \frac{1}{M!}(\mathcal{L}_{\boldsymbol{f}}^M \varphi)(t - \tau)^M + O((t - \tau)^{M+1}) \tag{6}$$

We assume the condition (4) does not hold, i.e. the following formula holds.

$$\forall \boldsymbol{x} \in H : \forall N = 2k - 1 : \bigvee_{j=1}^{N-1} \mathcal{L}_{\boldsymbol{f}}^{j}\varphi \neq 0 \vee \mathcal{L}_{\boldsymbol{f}}^{N}\varphi = 0 \qquad (7)$$

If $M = +\infty$, according to the formula (6), we have $\forall t \in (\tau - \epsilon, \tau + \epsilon) :$ $\varphi(\boldsymbol{x}(t)) = 0$, which contradicts the formula (3). If $M < \infty$, then,

1. if M is an even number, according to equation (6), we have $\forall t \in (\tau - \epsilon, \tau + \epsilon)\backslash\{\tau\} : \varphi(\boldsymbol{x}(t)) > 0$ when $\mathcal{L}_{\boldsymbol{f}}^{M} f_i > 0$, or $\forall t \in (\tau - \epsilon, \tau + \epsilon)\backslash\{\tau\} : \varphi(\boldsymbol{x}(t)) < 0$ when $\mathcal{L}_{\boldsymbol{f}}^{M} f_i < 0$, which contradicts the condition (3), or
2. if M is an odd number, it contradicts the condition (7).

Therefore, we have that (3) $\Rightarrow$ (4) holds.   □

Remark 1. The Formula (4) in Theorem 2 is a sufficient and necessary condition for deciding whether a system can pass through a hypersuface defined by a multivariate polynomial. In practice, univariate polynomials (i.e. $x_i - c$) instead of multivariate ones are most widely used for partitioning for their simplicity, where the resulting partition is rectangular, hence, Theorem 2 applies naturally. Note that in this simplified case,

- if $N = 1$, then Formula (4) simplifies to $\exists \boldsymbol{x}_\tau \in H : f_i(\boldsymbol{x}_\tau) \neq 0$ (where f_i is the i'th component of the vector flow function $\boldsymbol{f}$), which is used by Proposition 3 in [8]. This is the most intuitive way for a trajectory to pass through the hyperplane.
- if $N > 1$, which means $f_i(p) = 0$ and the vector field of a system is tangent to the hyperplane at $\boldsymbol{x}_\tau \in H$, the system is still capable of crossing H for having one of the odd-order Lie derivatives of $(x_i - c)$ be positive and all the other lower-order Lie derivatives vanish at $\boldsymbol{x}_\tau$.

In Theorem 2, higher-order Lie derivatives are used to characterize the relationship between a hypersurface and a vector flow. In fact, there are also other theories on hybrid systems which are based on higher-order Lie derivatives [24–26]. In [27], J. Liu et al. used higher-order Lie derivatives to describe a necessary and sufficient condition for a multivariate polynomial to be an inductive invariant for a continuous system, which needs to check all positive integers for the existence of a positive integer $N > 0$ such that the N'th-order Lie derivative is negative and all the i'th-order Lie derivative (for $i < N$) are equal to 0. In our case, we only need to check the existence of an odd number N such that the N'th-order Lie derivative is positive or negative, depending on the direction of the trajectory that we want to check at the boundary of $\{\boldsymbol{x} \in \mathbb{R} \mid \varphi(\boldsymbol{x}) \leq 0\}$ and all the other lower-order Lie derivatives are 0. Note that the $\varphi(\boldsymbol{x})$'s are not limited to univariate polynomials, that is, our partitions are not limited to rectangular regions.

One key point to apply Formula (4) is how to determine the constant N. Fortunately, there exists a computable upper bound M for N based on Gröbner basis theory [27]. Since the continuous systems under consideration are assumed to be multiaffine, $\mathcal{L}_f^j \varphi$ must be a polynomial in $\mathbb{R}[x]$. According to Theorem 1, we have

$$N \leq \gamma = \min \left\{ j \mid \mathcal{L}_f^{j+1} \varphi \in \langle \mathcal{L}_f^0 \varphi, ..., \mathcal{L}_f^j \varphi \rangle \right\} \tag{8}$$

The principle for Formula (8) is trivial, since for every $k \geq \gamma$ there must exist some $h_0, ..., h_r \in \mathbb{R}[x]$ such that $\mathcal{L}_f^k \varphi = \sum_{r=0}^{\gamma} h_r \mathcal{L}_f^r \varphi$. If $N > \gamma$, $\mathcal{L}_f^k \varphi$ must be 0 for all $k \geq 0$, which contradicts the fact that N satisfies Formula (4). The value of γ is computed iteratively by using the Gröbner basis. To compute the Gröbner basis, some powerful tool packages are available in popular mathematical softwares such as *Maple*. We implemented Algorithm 1 in Maple to compute γ. In Algorithm 1, R represents the high-order Lie derivative of φ, the function *GrobnerBasis* is used to compute the Gröbner basis G of $\langle \mathcal{L}_f^0 \varphi, ..., \mathcal{L}_f^{\gamma-1} \varphi \rangle$, and the function *NormForm* is used to compute the remainder of a polynomial w.r.t. a Gröbner basis. The iteration terminates if and when the remainder U is 0, which means $\mathcal{L}_f^\gamma \varphi \in \langle G \rangle$.

Algorithm 1. Compute the constant γ for polynomial φ.

Data: $f = [f_1, ..., f_n]$, φ
Result: γ
1 $R \leftarrow \varphi$;
2 $\gamma \leftarrow 0$;
3 $U \leftarrow R$;
4 $B \leftarrow \{R\}$;
5 **while** $U \neq 0$ **do**
6     $G \leftarrow GrobnerBasis(B)$;
7     $R \leftarrow \sum_{i=1}^{n} \frac{\partial R}{\partial x_i} f_i$;
8     $U \leftarrow NormForm(R, G)$;
9     $B \leftarrow B \cup \{R\}$;
10    $\gamma \leftarrow \gamma + 1$;
11 **end**

Based on Theorem 2, we further derive the following corollary for establishing a discrete transition between adjacent abstract states.

Corollary 1. *Given a continuous system $C = \langle X, f, Init \rangle$ and a set $\Phi = \{\varphi_i(x) \in \mathbb{R}[x], i = 1, \cdots, K\}$ of real coefficient polynomials, let $C_\Phi = \langle V_\Phi, T_\Phi, Init_\Phi \rangle$ be the corresponding discrete abstraction, where V_Φ is the abstract state space and $T_\Phi \in 2^{V_\Phi \times V_\Phi}$ is the set of abstract transitions, and let $u, v \in V_\Phi$. Then there exists a discrete transition $e = (u, v) \in T_\Phi$ if and only if $Adj(u, v)$ and*

$$\exists x \in Con\left(Min\left(u, v\right)\right) : \forall i = 1, \cdots, K : u_i \neq v_i \implies$$

$$\exists N_i = 2m_i - 1 : \bigwedge_{j=1}^{N_i - 1} \mathcal{L}_f^j \varphi_i = 0 \bigwedge (v_i - u_i) \mathcal{L}_f^{N_i} \varphi_i > 0 \tag{9}$$

where Min $(\boldsymbol{u}, \boldsymbol{v})$ returns the state of lower dimension and $m_i \in \mathbb{N}$.

Proof. By Theorem 2, we can easily prove that the corollary holds.

Remark 2. Here we give an intuitive explanation for Corollary 1. According to Definition 4, if there is a transition from $\boldsymbol{u}$ to $\boldsymbol{v}$, it must be one of the following two cases: (1) $\dim(\boldsymbol{u}) > \dim(\boldsymbol{v})$, then there exists a trajectory which reaches the intersection of the hypersurfaces $\{\varphi_i = 0 \mid u_i \neq v_i, i = 1, \cdots, K\}$, or (2) $\dim(\boldsymbol{u}) < \dim(\boldsymbol{v})$, then there exists a trajectory which escapes the intersection of the hypersurfaces. However, no matter in which case, we can decide the direction of the trajectory only by the signs of the higher-order Lie derivatives of $\{\varphi_i \mid u_i \neq v_i, i = 1, \cdots, K\}$ in the domain $Con(Min(\boldsymbol{u}, \boldsymbol{v}))$. Moreover, suppose $\dim(\boldsymbol{u}) > \dim(\boldsymbol{v})$; if there is a transition $(\boldsymbol{u}, \boldsymbol{v})$ or $(\boldsymbol{v}, \boldsymbol{u})$, it is easy to show that there is also a transition $(\boldsymbol{v}, \text{-}\boldsymbol{u})$ or $(\text{-}\boldsymbol{u}, \boldsymbol{v})$ correspondingly.

Now, we use an example to demonstrate the application of Corollary 2 to establishing discrete transitions.

Example 1. Consider the following multiaffine system.

$$
\begin{bmatrix} \dot{x}_1 \\ \dot{x}_2 \\ \dot{x}_3 \end{bmatrix} = \begin{bmatrix} -\frac{4(-x_2 x_3 + 12)}{13} - 2x_1 \\ 4(12 - x_1) - x_2 \\ 6(12 - x_1) - 4x_3 \end{bmatrix}
$$

Let $\Phi = \{\varphi_1(\boldsymbol{x}) = x_1 - 8, \varphi_2(\boldsymbol{x}) = x_2 - 8, \varphi_3(\boldsymbol{x}) = x_3 - 8\}$ be the set of polynomials for partitioning. We aim to decide the transition relation between the abstract states $\boldsymbol{u} = (-1, -1, 1)$ and $\boldsymbol{v} = (0, 0, 0)$, which corresponds to the regions $R_{\boldsymbol{u}} = \{\boldsymbol{x} \in \mathbb{R}^3 \mid \varphi_1(\boldsymbol{x}) < 0, \varphi_2(\boldsymbol{x}) < 0, \varphi_3(\boldsymbol{x}) > 0\}$ and $R_{\boldsymbol{v}} = \{\boldsymbol{x} \in \mathbb{R}^3 \mid \varphi_i(\boldsymbol{x}) = 0, i = 1, 2, 3\}$, respectively.

By applying Algorithm 1, we can get $\gamma_1 = 4$, $\gamma_2 = \gamma_3 = 3$. Since $\dim(\boldsymbol{u}) = 3 > 0 = \dim(\boldsymbol{v})$ and $Con(\boldsymbol{v})$ contains a single point $\boldsymbol{x}_\tau = (8, 8, 8)$, by computing the Lie derivatives of $\varphi_i(\boldsymbol{x})$ at $\boldsymbol{x}_\tau$, we get

$$
\mathcal{L}_f \varphi_1|_{\boldsymbol{x}=\boldsymbol{x}_\tau} = 0, \mathcal{L}_f^2 \varphi_1|_{\boldsymbol{x}=\boldsymbol{x}_\tau} = 0, \mathcal{L}_f^3 \varphi_1|_{\boldsymbol{x}=\boldsymbol{x}_\tau} = \frac{256}{13},
$$

$$
\mathcal{L}_f \varphi_2|_{\boldsymbol{x}=\boldsymbol{x}_\tau} = 8, \mathcal{L}_f \varphi_3|_{\boldsymbol{x}=\boldsymbol{x}_\tau} = -8
$$

In this case, as shown in Fig. 1, the trajectory of the system is tangent to the plane $x_1 - 8 = 0$ at $\boldsymbol{x}_\tau$. Since $\mathcal{L}_f \varphi_1 = 0$ at $\boldsymbol{x}_\tau$, we cannot decide the direction of the transition between $\boldsymbol{u}$ and $\boldsymbol{v}$ by considering only the first-order Lie derivative like in [10]. However, according to Corollary 1, we can decide that there is a discrete transition from $\boldsymbol{u}$ to $\boldsymbol{v}$.

4.2   Computation Method

In the previous section, we have introduced a necessary and sufficient condition for deciding whether there exists a discrete transition between two abstract

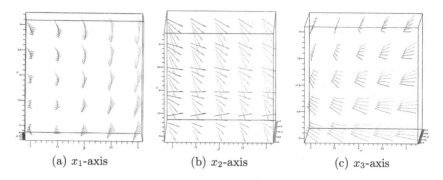

|(a) x_1-axis|(b) x_2-axis|(c) x_3-axis|

Fig. 1. Observing the vector fields of Example 1 from 3 different perspectives of view. The vector field is tangent to the plane $x_1 = 8$ at $\boldsymbol{x}_\tau = (8, 8, 8)$.

states. According to Corollary 1, the essential problem is to decide whether Formula (9) holds for two adjacent states $\boldsymbol{u}$ and $\boldsymbol{v}$, which involves deciding the satisfiability of a first-order logic formula built on a set of nonlinear polynomial equations and inequalities. To solve this problem, a number of SMT solvers are available, such as *Z3*, *SMT-RAT* etc.

Given two abstract states $\boldsymbol{u}$ and $\boldsymbol{v}$, we use Algorithm 2 to decide the transition relation between them. The idea of the algorithm is as follows. Given any two abstract states $\boldsymbol{u}$ and $\boldsymbol{v}$, if they are adjacent, we select the one with lower dimension, represented as $\boldsymbol{w} = Min\,(\boldsymbol{u}, \boldsymbol{v})$. As commented in Remark 2, the higher-order Lie derivatives of $\{\varphi_i(x) \mid u_i \neq v_i, i = 1, \cdots, K\}$ in the domain $Con\,(Min\,(\boldsymbol{u}, \boldsymbol{v}))$ suffice to determine the transition relation between $\boldsymbol{u}$ and $\boldsymbol{v}$. In Algorithm 2, the set $I_{\boldsymbol{u},\boldsymbol{v}}$ collects the indices of the hypersurfaces where $\boldsymbol{u}$ differs from $\boldsymbol{v}$. Lines 5 to 8 are used to construct all the possible combinations of Lie derivatives of $\{\varphi_i \mid u_i \neq v_i, i = 1, \cdots K\}$ occurring in Formula (9). Then, we check with an SMT solver in line 10 if there exists a combination of the Lie derivatives which makes Formula (9) hold.

5   Discussion and Examples

In this section, we analyze and demonstrate the improvement of our method over the existing methods using some examples. Note that to be intuitive, we refer an abstract state to its region of concrete states in the following.

Theorem 2 presents a sufficient and necessary condition for deciding the existence of a trajectory passing through a hypersurface $\varphi(\boldsymbol{x}) = 0$ at a single point by a series of Lie derivatives of $\varphi(\boldsymbol{x})$. However, as indicated in Corollary 1, what we need to handle are mainly semialgebraic sets which are usually infinite. For example, given a rectangular partition for a 3-dimensional state space, the adjacent regions of a box consist of rectangles, edges, and vertices. To decide whether there exists a trajectory passing through an adjacent region, a general solution is to apply an SMT solver, which is known to have a doubly-exponential complexity. Hence, there are other attempts to simplify this problem in special cases.

Algorithm 2. Decide the direction of transition between $\boldsymbol{u}$ and $\boldsymbol{v}$.

Data: abstract states $\boldsymbol{u}, \boldsymbol{v}$; polynomial set $\Phi = \{\varphi_i, i = 1, \cdots, K\}$;
the constant array $\Gamma = [\gamma_1, \cdots, \gamma_K]$ for all $\varphi_i \in \Phi$

Result: $\boldsymbol{e}$: the transition relation

1 **if** $\boldsymbol{u}$ *and* $\boldsymbol{v}$ *are adjacent* **then**

2 $\quad$ select the lower-dimensional state $\boldsymbol{w} = Min\,(\boldsymbol{u}, \boldsymbol{v})$;

3 $\quad$ $P_w \leftarrow$ set of polynomial predicates defining $Con\,(\boldsymbol{w})$;

4 $\quad$ $I_{u,v} \leftarrow \{i \mid u_i \neq v_i, i = 1, \cdots, K\}$;

5 $\quad$ **for** *each* $i \in I_{u,v}$ **do**

6 $\quad\quad$ $\Psi_i \leftarrow \{\bigwedge_{j=1}^{N_i-1} \mathcal{L}_f^j \varphi_i = 0 \bigwedge (v_i - u_i)\mathcal{L}_f^{N_i} \varphi_i > 0 \mid 1 \leq N_i \leq \gamma_i, N_i \text{ is odd}\}$;

7 $\quad$ **end**

8 $\quad$ $\Psi \leftarrow \{(\psi_1, ..., \psi_M) \mid \psi_j \in \Psi_{i_j}, i_1 < \cdots < i_M, M = |I_{u,v}|, i_j \in I_{u,v}\}$;

9 $\quad$ **for** *each* $(\psi_1, \cdots, \psi_M) \in \Psi$ **do**

10 $\quad\quad$ **if** $(\bigwedge_{p_i \in P_w} p_i) \wedge (\bigwedge_{j=1}^{M} \psi_j)$ *is satisfiable by SMT solver* **then**

11 $\quad\quad\quad$ $\boldsymbol{e} \leftarrow (\boldsymbol{u}, \boldsymbol{v})$;

12 $\quad\quad\quad$ break;

13 $\quad\quad$ **end**

14 $\quad$ **end**

15 $\quad$ $\boldsymbol{e} \leftarrow 0$;

16 **end**

In [8], G. Batt et al. presented a sufficient and necessary condition for deciding the existence of transitions between adjacent hyperrectangles (their definition of *adjacent* regions refers to those n-dimensional hyperrectangles having an $(n-1)$-dimensional facet in common). The condition states that for any two adjacent full-dimensional hyperrectangles R and R' (assuming R' is greater than R in x_i), there exists a transition from R to R' if and only if there exists a vertex on the shared facet F_i of R and R' satisfying $\dot{x}_i = f_i > 0$ (note that this is not true for adjacent hyperrectangles with shared facets of dimension lower than $n-1$). The condition is obvious for the sufficiency, but not for the necessity according to Theorem 2. To prove the necessity, we need to prove that there exists no trajectory from R to R' if $f_i \leq 0$ for all the vertices of F_i. This can be addressed in two cases:

- $f_i < 0$ for some vertex. According to the property of multiaffine functions described in Proposition 1, it is obvious that $f_i(x_\tau) < 0$ for all $x_\tau \in F_i \backslash \partial F_i$, where ∂F_i denotes the boundary of f_i. Therefore, all the trajectories that pass through F_i at the internal points must point from R' to R instead of the reverse.

- $f_i = 0$ for all vertices. We can easily derive from Proposition 1 that $f_i = 0$ for all the internal points of F_i. However, according to Theorem 2, we cannot conclude that there exist no trajectories from R to R' without further knowledge about the higher-order derivatives of $(x_i - c)$. Nevertheless, we have the following proposition which asserts that all the Lie derivatives of $(x_i - c)$ at any internal points of F_i are 0 if the first Lie derivative of x_i at all the vertices of F_i are 0. Therefore, there exists no trajectory no matter in which direction, i.e. from R to R' or reversely.

Proposition 2. *Given a multiaffine system $\dot{x} = f(x)$ and a rectangular state space partition, where $f(x) = (f_1(x), ..., f_n(x))$, let R and R' be two n-dimensional hyperrectangles which share an $(n-1)$-dimensional facet F_i in the partition, where $F_i = \{(x_1, ..., x_n) \in \mathbb{R}^n \mid x_i = c, x_j \in [a_j, b_j], a_j \in \mathbb{R}, b_j \in \mathbb{R}, 1 \le j \le n, j \ne i\}$, and let $\mathbb{V}(F_i)$ denote the set of vertices of F_i. Then the following formula holds:*

$$\forall x \in \mathbb{V}(F_i) : f_i(x) = 0 \implies \forall x \in F_i : \forall M \ge 0 : \mathcal{L}_f^M f_i = 0 \tag{10}$$

Moreover, $x_i - c = 0$ is an invariant by the right hand side of Formula (10).

Proof. Suppose $\forall x \in \mathbb{V}(F_i) : f_i(x) = 0$ holds. According to Proposition 1, we can easily derive that $\forall x \in F_i : f_i(x) = 0$, which means that $f_i(x)$ must be of the form $f_i(x) = (x_i - c)P_1(x)$, where $P_1(x)$ is a multiaffine function in $\mathbb{R}[x_1, ..., x_{i-1}, x_{i+1}, ..., x_n]$. To prove that $\forall x \in F_i : \forall M \ge 0 : \mathcal{L}_f^M f_i = 0$ holds, we only need to prove that every $\mathcal{L}_f^M f_i$ has the form of $(x_i - c)P_M(x)$, where $P_M(x) \in \mathbb{R}[x]$. By induction, we assume that $\mathcal{L}_f^{M-1} f_i = (x_i - c)P_{M-1}(x)$, so we only need to prove that $\mathcal{L}_f^M f_i = (x_i - c)P_M(x)$. We have the following equation.

$$\mathcal{L}_f^M f_i = \mathcal{L}_f(\mathcal{L}_f^{M-1} f_i) = (x_i - c)\mathcal{L}_f P_{M-1} + P_{M-1}\mathcal{L}_f(x_i - c)$$
$$= (x_i - c)\mathcal{L}_f P_{M-1} + P_{M-1}(x_i - c)P_1 = (x_i - c)(\mathcal{L}_f P_{M-1} + P_{M-1}P_1) \tag{11}$$

With $P_M(x) = \mathcal{L}_f P_{M-1} + P_{M-1}P_1$, the above equation can be written as

$$\mathcal{L}_f^M f_i = (x_i - c)P_M(x) \tag{12}$$

Therefore, we can conclude that Formula (10) holds. Moreover, by the Taylor expansion of $x_i(t)$ at $x_i = c$ we can easily prove that $(x_i - c)$ is an invariant. $\square$

Proposition 2 shows that there exists no trajectory connecting R and R' when $f_i = 0$ at all the vertices of F_i. However, this does not mean that there exists no trajectory that can reach F_i. In fact, there could exist an infinite number of trajectories which can reach F_i in the hyperplane P_{F_i} containing F_i. In fact, P_{F_i} forms an invariant of the system state space, i.e. any trajectory starting from P_{F_i} will never escape from P_{F_i}. To construct a precise over-approximation for the original system, these trajectories are non-negligible. However, these trajectories cannot be handled by the abstraction method in [8] because the authors only consider the transitions between adjacent full-dimensional hyperrectangles but ignore the transitions between the lower-dimensional hyperrectangles. In the following, we present an example to demonstrate the case.

Example 2. Consider the following multiaffine system.

$$\begin{bmatrix} \dot{x}_1 \\ \dot{x}_2 \\ \dot{x}_3 \end{bmatrix} = \begin{bmatrix} f_1(x) \\ f_2(x) \\ f_2(x) \end{bmatrix} = \begin{bmatrix} x_2x_3 + x_1 + 1 \\ x_3 + x_1 + 1 \\ (x_3 - 1)(x_1x_2 + 1) \end{bmatrix} \tag{13}$$

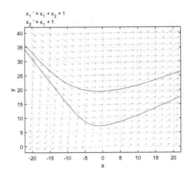

$x_1' = x_1 + x_2 + 1$
$x_2' = x_1 + 1$

Fig. 2. Section of the vector field of Example 2 on the plane $x_3 = 1$.

Let $Init = \{(x_1, x_2, x_3) \in \mathbb{R}^3 \mid (x_1, x_2, x_3) \in [-10, 15] \times [15, 20] \times [0, 2]\}$ be the initial set and $Uns = \{(x_1, x_2, x_3) \in \mathbb{R}^3 \mid (x_1, x_2, x_3) \in [15, 20] \times [20, 25] \times [1, 1]\}$ be the unsafe set.

Suppose we have a rectangular partition for the system in Example 2 that contains two boxes R and R' which share the facet F_i, where $\mathbb{V}(F_i) = \{(0, 0, 1), (1, 0, 1), (0, 1, 1), (1, 1, 1)\}$. We can easily verify that $\mathcal{L}_f^M f_3(\boldsymbol{x}) = 0$ on the plane $x_3 = 1$ for all $M \geq 0$, which means that all the pairs of boxes besides (R, R') that share the plane $x_3 = 1$ cannot reach one another. However, according to Proposition 2, $x_3 = 1$ is an invariant and there could exist plenty of trajectories on the plane $x_3 = 1$ for which only the variables x_1 and x_2 change over time while x_3 stays unchanged. The section of the vector field at the plane $x_3 = 1$ is shown in Fig. 2, and we can see that the system is unsafe. However, by using the abstraction method in [8], the transitions in the plane $x_3 = 1$ are ignored and hence the resulting transition system could be verified to be safe.

In [23], M. Kloetzer and C. Belta constructed an over-approximation for the original system by taking into account all the transitions between the hyper-rectangles of different dimensions from 0 to n. In order to deal with the case of trajectories being tangent to the shared facet of neighboring hyperrectangles (i.e. $f_i(\boldsymbol{x}) = 0$), the authors decide the direction of the trajectory by the direction of the vector flow in the neighboring hyperrectangles. This strategy works well only when the direction of the vector flow in the neighboring region is definite. Otherwise, the procedure fails and bidirectional transitions between two hyper-rectangles have to be added in order to get an over-approximation. We present two examples to demonstrate how this can be handled by our method.

Example 3. Consider the following 3-dimensional multiaffine system.

$$\begin{bmatrix} \dot{x}_1 \\ \dot{x}_2 \\ \dot{x}_3 \end{bmatrix} = \begin{bmatrix} f_1(\boldsymbol{x}) \\ f_2(\boldsymbol{x}) \\ f_3(\boldsymbol{x}) \end{bmatrix} = \begin{bmatrix} x_1 - 1 \\ x_2 + 1 \\ x_1 x_2 + x_1 + x_2 \end{bmatrix} \tag{14}$$

Let $\{-1, 0, 1\}^3$ be the set of grid points of a rectangular partition. What is the transition starting from the origin $(0, 0, 0)$?

For the system in Example 3, the flow vector at $(0,0,0)$ is $(-1,1,0)$. Since $\dot{x}_3 = 0$, to decide the target hyperrectangle of $(0,0,0)$ according to the algorithm in [23], the authors have to decide the direction of the vector flow in $H_0 = \{(x_1, x_2, x_3) \in \mathbb{R}^3 \mid -1 < x_1 < 0, 0 < x_2 < 1, x_3 = 0\}$. Unfortunately, the direction of the vector flow is indefinite in H_0, and so they have to add two transitions: $(0,0,0) \to H_1$ and $H_0 \to (0,0,0)$, which is apparently not reasonable because there could exist only one trajectory passing through the origin. However, by using Corollary 1, we can easily decide that the trajectory enters the region of $\{(x_1, x_2, x_3) \in \mathbb{R}^3 \mid -1 < x_1 < 0, 0 < x_2 < 1, -1 < x_3 < 0\}$ by $\mathcal{L}_f x_1 = -1$, $\mathcal{L}_f x_2 = 1$, $\mathcal{L}_f x_3 = \mathcal{L}_f^2 x_3 = 0$ and $\mathcal{L}_f^3 x_3 = -2$.

Example 4. Consider the following 2-dimensional multiaffine system.

$$\begin{bmatrix} \dot{x}_1 \\ \dot{x}_2 \end{bmatrix} = \begin{bmatrix} f_1(x) \\ f_2(x) \end{bmatrix} = \begin{bmatrix} (1-x_1)x_2 \\ x_1 + x_2 + 1 \end{bmatrix} \tag{15}$$

Let $\{0,1,2\} \times \{0,1,2\}$ be the set of grid points of a rectangular partition. What are the transitions starting from the region $H_1 = \{(x_1, x_2) \in \mathbb{R}^2 \mid 0 < x_1 < 1, 0 < x_2 < 1\}$?

For the system in Example 4, it is easy to verify that $\dot{x}_1 > 0$ and $\dot{x}_2 > 0$ for all $(x_1, x_2) \in H_1$. Let $H_2 = \{(x_1, x_2) \in \mathbb{R}^2 \mid x_1 = 1, 0 < x_2 < 1\}$, $H_3 = \{(x_1, x_2) \in \mathbb{R}^2 \mid x_1 = x_2 = 1\}$ and $H_4 = \{(x_1, x_2) \in \mathbb{R}^2 \mid 0 < x_1 < 1, x_2 = 1\}$. According to the algorithm in [23], there should be the following transitions in the abstraction: $H_1 \to H_1$, $H_1 \to H_2$, $H_1 \to H_3$ and $H_1 \to H_4$. According to Corollary 1, however, there could not exist transitions from H_1 to H_2 and H_3 due to the fact that $L_f^m(x_1 - 1) = 0$ for all $m > 0$ in $L_1 = \{(x_1, x_2) \in \mathbb{R}^2 \mid x_1 = 1\}$. In fact, L_1 is an invariant, which means that any trajectory reaching L_1 must start from and always stay in L_1.

Recall that Proposition 2 concludes that no trajectory can pass through an $(n-1)$-dimensional facet F_i in an n-dimensional space if $f_i(v_j) = 0$ for every vertex v_j of F_i, where f_i is the i'th component of the vector flow f. However, this is not true for a facet F_i of lower dimension than $n - 1$. In other words, there could exist trajectories that can pass though an edge in the direction of x_i even if $f_i(x) = 0$ for all x in F_i. Let us demonstrate this case using the following example.

Example 5. Consider the following 4-dimensional multiaffine system.

$$\begin{bmatrix} \dot{x}_1 \\ \dot{x}_2 \\ \dot{x}_3 \\ \dot{x}_4 \end{bmatrix} = \begin{bmatrix} f_1(x) \\ f_2(x) \\ f_3(x) \\ f_4(x) \end{bmatrix} = \begin{bmatrix} -x_4 x_2 + x_4 x_3 + 2x_2 x_3 + x_1 - x_2 - 3x_3 + 1 \\ x_3 \\ x_1 \\ x_1 + x_2 \end{bmatrix} \tag{16}$$

Let $(1,1,1,3)$ and $(1,1,1,4)$ be the grid points of a rectangular partition, $R_1 = \{(x_1, x_2, x_3, x_4) \in \mathbb{R}^4 \mid x_1 < 1, x_2 = 1, x_3 = 1, 3 < x_4 < 4\}$, $R_2 = \{(x_1, x_2, x_3, x_4) \in \mathbb{R}^4 \mid x_1 = x_2 = x_3 = 1, 3 < x_4 < 4\}$ and

$R_3 = \{(x_1, x_2, x_3, x_4) \in \mathbb{R}^4 \mid x_1 > 1,\ x_2 = 1, x_3 = 1, 3 < x_4 < 4\}$. We know that both R_1 and R_3 are adjacent to R_2. To establish the transitions between them, we need to check the signs of the Lie derivatives of $x_1 - 1$ in R_2. Let $L_1 = \langle x_1 - 1, x_2 - 1, x_3 - 1 \rangle$. We can verify that $\mathcal{L}_f(x_1 - 1), \mathcal{L}_f^2(x_1 - 1) \in L_1$ and the remainder of $\mathcal{L}_f^3(x_1 - 1)$ w.r.t. L_1 is $(5 - x_4)$, which means that in R_2 both $\mathcal{L}_f(x_1 - 1)$ and $\mathcal{L}_f^2(x_1 - 1)$ are identical to 0 and $\mathcal{L}_f^3(x_1 - 1) > 0$. Hence, according to Corollary 1, there are transitions $R_1 \rightarrow R_2$ and $R_2 \rightarrow R_3$. Obviously, this is not decidable only by first-order Lie derivatives as in [8, 23].

6   Conclusion

In this paper, in order to reduce the conservativeness of the discrete abstraction, we proposed a new method based on a sufficient and necessary decision condition for establishing the discrete transitions between the abstract states in the abstract system. The partition of the state space of a multiaffine system is assumed to be based on a set of multivariate polynomials. A rectangular partition is just a simple case of a multivariate polynomial partition and the new decision condition applies naturally. Examples show the improvement of our method over the existing methods.

Acknowledgments. This research was supported in part by the Austrian Science Fund (FWF) under grants S11402-N23, S11405-N23 and S11412-N23 (RiSE/SHiNE) and Z211-N23 (Wittgenstein Award).

References

1. Alur, R., Dang, T., Ivančić, F.: Progress on reachability analysis of hybrid systems using predicate abstraction. In: Maler, O., Pnueli, A. (eds.) HSCC 2003. LNCS, vol. 2623, pp. 4–19. Springer, Heidelberg (2003)
2. Asarin, E., Dang, T., Girard, A.: Reachability analysis of nonlinear systems using conservative approximation. In: Maler, O., Pnueli, A. (eds.) HSCC 2003. LNCS, vol. 2623, pp. 20–35. Springer, Heidelberg (2003)
3. Bak, S., Bogomolov, S., Henzinger, T.A., Johnson, T.T., Prakash, P.: Scalable static hybridization methods for analysis of nonlinear systems. In: Proceedings of the 19th International Conference on Hybrid Systems: Computation and Control, HSCC 2016, Vienna, Austria, pp. 155–164, 12–14 April 2016
4. Bartocci, E., Corradini, F., Di Berardini, M.R., Entcheva, E., Smolka, S.A., Grosu, R.: Modeling and simulation of cardiac tissue using hybrid I/O automata. Theor. Comput. Sci. **410**(33), 3149–3165 (2009)
5. Bartocci, E., Corradini, F., Entcheva, E., Grosu, R., Smolka, S.A.: Cellexcite: an efficient simulation environment for excitable cells. BMC Bioinform. **9**, S–2 (2008)
6. Bartocci, E., Lió, P.: Computational modeling, formal analysis and tools for systems biology. PLOS Comput. Biol. **12**(1), e1004591 (2016)
7. Bartocci, E., Liò, P., Merelli, E., Paoletti, N.: Multiple verification in complex biological systems: the bone remodelling case study. Trans. Comput. Syst. Biol. **14**, 53–76 (2012)

8. Batt, G., Belta, C., Weiss, R.: Temporal logic analysis of gene networks under parameter uncertainty. Trans. Autom. Control **53**(Special Issue), 215–229 (2008)
9. Batt, G., De Jong, H., Page, M., Geiselmann, J.: Symbolic reachability analysis of genetic regulatory networks using discrete abstractions. Automatica **44**(4), 982–989 (2008)
10. Belta, C., Habets, L.C.: Controlling a class of nonlinear systems on rectangles. Trans. Autom. Control **51**(11), 1749–1759 (2006)
11. Bogomolov, S., Donzé, A., Frehse, G., Grosu, R., Johnson, T.T., Ladan, H., Podelski, A., Wehrle, M.: Guided search for hybrid systems based on coarse-grained space abstractions. Int. J. Softw. Tools Technol. Transf. **18**(4), 449–467 (2016). doi:10.1007/s10009-015-0393-y
12. Bogomolov, S., Frehse, G., Greitschus, M., Grosu, R., Pasareanu, C., Podelski, A., Strump, T.: Assume-guarantee abstraction refinement meets hybrid systems. In: Yahav, E. (ed.) HVC 2014. LNCS, vol. 8855, pp. 116–131. Springer, Heidelberg (2014)
13. Bogomolov, S., Schilling, C., Bartocci, E., Batt, G., Kong, H., Grosu, R.: Abstraction-based parameter synthesis for multiaffine systems. In: Piterman, N. (ed.) HVC 2015. LNCS, vol. 9434, pp. 19–35. Springer, Heidelberg (2015). doi:10.1007/978-3-319-26287-1_2
14. Cox, D.A., Little, J., O'Shea, D.: Ideals, Varieties and Algorithms: An Introduction to Computational Algebraic Geometry and Commutative Algebra. Springer, New York (2007)
15. Dang, T., Le Guernic, C., Maler, O.: Computing reachable states for nonlinear biological models. In: Degano, P., Gorrieri, R. (eds.) CMSB 2009. LNCS, vol. 5688, pp. 126–141. Springer, Heidelberg (2009)
16. Dreossi, T., Dang, T.: Parameter synthesis for polynomial biological models. In: Proceedings of the 17th International Conference on Hybrid Systems: Computation and Control, pp. 233–242. ACM (2014)
17. Grosu, R., Batt, G., Fenton, F.H., Glimm, J., Le Guernic, C., Smolka, S.A., Bartocci, E.: From cardiac cells to genetic regulatory networks. In: Gopalakrishnan, G., Qadeer, S. (eds.) CAV 2011. LNCS, vol. 6806, pp. 396–411. Springer, Heidelberg (2011)
18. Habets, L., Collins, P.J., van Schuppen, J.H.: Reachability and control synthesis for piecewise-affine hybrid systems on simplices. IEEE Trans. Autom. Control **51**(6), 938–948 (2006)
19. Habets, L., Van Schuppen, J.H.: A control problem for affine dynamical systems on a full-dimensional polytope. Automatica **40**(1), 21–35 (2004)
20. Habets, L.C.G.J.M., Van Schuppen, J.: A control problem for affine dynamical systems on a full-dimensional simplex. Rep. Probab. Netw. Algorithms **17**, 1–19 (2000)
21. Jiang, Y., Zhang, H., Li, Z., Deng, Y., Song, X., Gu, M., Sun, J.: Design and optimization of multiclocked embedded systems using formal techniques. IEEE Trans. Ind. Electron. **62**(2), 1270–1278 (2015)
22. Jiang, Y., Zhang, H., Zhang, H., Liu, H., Song, X., Gu, M., Sun, J.: Design of mixed synchronous/asynchronous systems with multiple clocks. IEEE Trans. Parallel Distrib. Syst. **26**(8), 2220–2232 (2015)
23. Kloetzer, M., Belta, C.: Reachability analysis of multi-affine systems. In: Hespanha, J.P., Tiwari, A. (eds.) HSCC 2006. LNCS, vol. 3927, pp. 348–362. Springer, Heidelberg (2006)
24. Kong, H., Bogomolov, S., Schilling, C., Jiang, Y., Henzinger, T.A., Invariant clusters for hybrid systems. arXiv preprint arXiv: 1605.01450 (2016)

25. Kong, H., He, F., Song, X., Hung, W.N.N., Gu, M.: Exponential-condition-based barrier certificate generation for safety verification of hybrid systems. In: Sharygina, N., Veith, H. (eds.) CAV 2013. LNCS, vol. 8044, pp. 242–257. Springer, Heidelberg (2013)
26. Kong, H., Song, X., Han, D., Gu, M., Sun, J.: A new barrier certificate for safety verification of hybrid systems. Comput. J. **57**(7), 1033–1045 (2014)
27. Liu, J., Zhan, N., Zhao, H.: Computing semi-algebraic invariants for polynomial dynamical systems. In: International Conference on Embedded Software, pp. 97–106. ACM (2011)
28. Sogokon, A., Ghorbal, K., Jackson, P.B., Platzer, A.: A method for invariant generation for polynomial continuous systems. In: Jobstmann, B., Leino, K.R.M. (eds.) VMCAI 2016. LNCS, vol. 9583, pp. 268–288. Springer, Heidelberg (2016). doi:10.1007/978-3-662-49122-5_13
29. Tiwari, A.: Abstractions for hybrid systems. Formal Methods Syst. Des. **32**(1), 57–83 (2008)
30. Tiwari, A., Khanna, G.: Series of abstractions for hybrid automata. In: Tomlin, C.J., Greenstreet, M.R. (eds.) HSCC 2002. LNCS, vol. 2289, pp. 465–478. Springer, Heidelberg (2002)

Stochastic Modelling

On Observability and Reconstruction of Promoter Activity Statistics from Reporter Protein Mean and Variance Profiles

Eugenio Cinquemani[(✉)]

Inria Grenoble – Rhône-Alpes, Montbonnot, 38334 St. Ismier Cedex, France
eugenio.cinquemani@inria.fr
https://team.inria.fr/ibis/eugenio-cinquemani/

Abstract. Reporter protein systems are widely used in biology for the indirect quantitative monitoring of gene expression activity over time. At the level of population averages, the relationship between the observed reporter concentration profile and gene promoter activity is established, and effective methods have been introduced to reconstruct this information from the data. At single-cell level, the relationship between population distribution time profiles and the statistics of promoter activation is still not fully investigated, and adequate reconstruction methods are lacking.

This paper develops new results for the reconstruction of promoter activity statistics from mean and variance profiles of a reporter protein. Based on stochastic modelling of gene expression dynamics, it discusses the observability of mean and autocovariance function of an arbitrary random binary promoter activity process. Mathematical relationships developed are explicit and nonparametric, i.e. free of a priori assumptions on the laws governing the promoter process, thus allowing for the decoupled analysis of the switching dynamics in a subsequent step. The results of this work constitute the essential tools for the development of promoter statistics and regulatory mechanism inference algorithms.

Keywords: Gene regulation · Doubly stochastic process · Spectral analysis

1   Introduction

A common experimental technique to monitor gene expression is the use of reporter proteins [9], i.e. fluorescent or luminescent proteins that are synthesized upon expression of the gene of interest. Light intensity measurements collected at different points in time are proportional to the amount of reporter molecules. This provides a quantitative, however indirect, readout of the activity of the gene, since reporter abundance depends on gene activation via its own transcription and translation dynamics.

When cellular populations are observed as a whole, such as in automated microplate readers, an average reporter profile is obtained. An estimate of the

© Springer International Publishing AG 2016
E. Cinquemani and A. Donzé (Eds.): HSB 2016, LNBI 9957, pp. 147–163, 2016.
DOI: 10.1007/978-3-319-47151-8_10

average gene activation over the population of cells may thus be obtained by regularized inversion of the reporter synthesis dynamics [21]. Provided accurate knowledge of the latter, reconstruction of the promoter activity allows one to investigate gene expression regulatory mechanisms, a crucial step toward inference of gene regulatory networks [18].

When individual cells are observed, for instance via flow-cytometry or fluorescence videomicroscopy, a statistical distribution of gene expression levels over a sample of the population (often called a population snapshot [6]) is obtained at several points in time (reporter traces for individual cells can also be obtained by suitable experimental setups and image processing techniques [20], but we will not analyze this case here). In many cases of interest, this crucially reveals variability of gene expression levels across cells that can be explained in terms of the stochasticity of the gene regulation and expression process [13,16,19]. Reporter statistics thus contains information about the stochastic laws governing gene activation. However, recovering the relevant information from the data is less trivial than in the population average case, and no satisfactory methods exist to date.

With reference to population snapshot data, in [2,3], we have started addressing the problem of estimating promoter activity statistics (the biological information of interest in gene expression reporting) from reporter mean and variance profiles. In [2], parametric models of stochastic gene activation have been considered, and the identifiability of promoter switching rates that are fixed over time and across cells has been analyzed. However, due to a priori unknown regulatory mechanisms, switching rates may fluctuate over time and/or across cells (extrinsic noise). To cope with this, in [3], a nonparametric method, i.e. avoiding assumptions on the regulatory mechanisms behind the expression of the gene of interest, has been proposed for the special case of irreversible activation. A rather extensive account of relevant research literature is also contained in these works.

Following up from the developments in [2,3], for the general case of unmodelled stochastic (possibly time varying) gene expression regulation, we address here the problem of reconstructing second-order statistics of the promoter activity process from reporter mean and variance profiles. The importance of this problem lies in the fact that, in analogy with linear stochastic processes [12], cross-correlation of promoter activity at different points in time (i.e. the auto-correlation function) contains information about the time dynamics of activation and deactivation. Reconstruction of these statistics from data is thus the crucial step for the understanding of the gene regulatory mechanisms at the level of single cells, where stochastic variability offers more to discover than traditional population analysis [13,14].

The contribution of this paper is the development of explicit relationships between the unknown (first- and) second-order promoter activity statistics and the experimentally measurable reporter mean and variance profiles. Crucially, these relationships rely on nonparametric models of gene activation, i.e. no a priori assumption is made except the absence of stochastic feedback from reporter abundance to the regulation of the gene itself, a hypothesis that agrees well with the biochemistry of reporter systems. Based on analytic investigation and examples, we show that these relationships are essentially linear, whence invertible in a tractable manner, and allow for the discrimination among different promoter activity regulatory laws. On these basis, the implementation of algorithms for the actual estimation of the statistics of interest is left for future work. For ease of reading, all mathematical proofs are deferred to Appendix A. Appendix B, instead, summarizes results from [2] that are used in this work.

2 Background Material

Gene expression monitoring over time is commonly operated by the use of fluorescent or luminescent reporter proteins (see [9] and references therein). In essence, synthesis of a reporter protein is placed under the control of the promoter of the gene of interest by engineering its coding sequence onto the DNA at an appropriate place. When the gene is expressed, transcription and subsequent translation leads to the formation of new reporter protein molecules. Whether luminescent or fluorescent, reporter protein molecules can be quantified at any time by measuring light intensity at the relevant wavelength, thus providing a dynamical readout of the activity of the gene. To do so, time-lapse microscopy, flow cytometry, microplate reading, or other experimental techniques are used, depending whether single-cell measurements, population histograms, or population-average profiles are sought. Synthesis of reporter proteins is often completed by a maturation step, that takes immature proteins into their mature, visible form.

2.1 Stochastic Gene Expression Modelling

Gene expression is commonly described in terms of the synthesis and degradation reactions for $mRNA$ and protein molecules

$$\mathscr{R}_1 : F \xrightarrow{k_M} F + M \qquad\qquad \mathscr{R}_2 : M \xrightarrow{d_M} \emptyset \qquad (1)$$

$$\mathscr{R}_3 : M \xrightarrow{k_P} M + P \qquad\qquad \mathscr{R}_4 : P \xrightarrow{d_P} \emptyset \qquad (2)$$

[5,10] where M and P denote $mRNA$ and protein species, respectively, and F represents the active promoter species. In the context of this paper, P is the fluorescent or luminescent reporter protein. We will not distinguish between immature (invisible) and mature (visible) protein molecules. If necessary (e.g. for slow, stochastic maturation), an additional first-order reaction $P \to P_{mature}$ can be included in the model (along with $P_{mature} \to \emptyset$) to account for protein maturation (and mature protein degradation).

Denote with $X_1 \in \mathbb{N}$ and $X_2 \in \mathbb{N}$ the number of copies of M and P, in the same order, and with $X_3 \in \{0, 1\}$ the state of the promoter, i.e. $X_3 = 0$ when the promoter is inactive (absence of F) and $X_3 = 1$ when it is active (presence of F). Switching promoter dynamics (responsible of $mRNA$ synthesis bursts in single cells) are formally captured by two additional reactions,

$$\mathscr{R}_5 : \emptyset \xrightarrow{\lambda_+ \cdot (1 - X_3)} F, \qquad\qquad \mathscr{R}_6 : F \xrightarrow{\lambda_-} \emptyset, \qquad (3)$$

representing in the order activation with propensity $\lambda_+ \cdot (1 - X_3)$ (only enabled if $X_3 = 0$), and deactivation with propensity $\lambda_- \cdot X_3$ (only enabled if $X_3 = 1$). Overall, this is a system of $m = 6$ chemical reactions over $n = 3$ different species.

The kinetics of this biochemical reaction system can be expressed in terms of stoichiometry matrix S and reaction rate vector $a(x)$ given by

$$
S = \begin{bmatrix} 1 & -1 & 0 & 0 & 0 & 0 \\ 0 & 0 & 1 & -1 & 0 & 0 \\ 0 & 0 & 0 & 0 & 1 & -1 \end{bmatrix}, \quad a(x) = \begin{bmatrix} k_M x_3 \\ d_M x_1 \\ k_P x_1 \\ d_P x_2 \\ \lambda_+(1 - x_3) \\ \lambda_- x_3 \end{bmatrix},
$$

where, for $i = 1, \ldots, n$ and $j = 1, \ldots, m$, $S_{i,j}$ denotes the net change in molecule number of species i when reaction $\mathcal{R}_j$ occurs. At the level of a single cell, $X = [X_1\ X_2\ X_3]^T$ is a stochastic process and, for every j, $a_j(x)$ is interpreted as the infinitesimal probability that reaction $\mathcal{R}_j$ occurs in an infinitesimal time period when $X = x$ molecules of the different species are present in the reaction volume [16]. For constant rates λ_+ and λ_-, Eqs. (1)–(3) together constitute the so-called random telegraph model [16]. In general, however, these rates might themselves depend upon the amount of transcription factors regulating the expression of the gene, which one may write as $\lambda_+(X_?)$ and $\lambda_-(X_?)$, with $X_?$ denoting the amount of some unspecified species.

The question we are going to investigate is what can be said about the statistics of F, given mean and variance profiles of the amounts of protein P across a population of cells. In practice, fluorescence or luminescence measurements proportional to the actual amount of protein are measured and are possibly affected by error. In this paper, however, we are not concerned with the details of the measurement model, and assume that mean and variance of X_2 are observed directly.

2.2   Propagation of Moments

Consider an arbitrary biochemical reaction system with n reactants, m reactions, stoichiometry matrix S and reaction rates $a(x, u)$ possibly depending on a deterministic input u. Let $X(t)$ be the corresponding random state vector at time t, and define $\mu(t) = \mathbb{E}[X(t)]$ and $\Sigma(t) = \mathrm{Cov}\big(X(t)\big) = \mathbb{E}\big[\big(X(t) - \mu(t)\big)\big(X(t) - \mu(t)\big)^T\big]$. It can be shown (see e.g. [8]) that μ and Σ obey the so-called moment equations

$$
\dot{\mu} = S\mathbb{E}[a(X, u)], \tag{4}
$$

$$
\dot{\Sigma} = S\mathbb{E}\big[a(X, u)(X - \mu)^T\big] + \mathbb{E}\big[(X - \mu)a^T(X, u)\big]S^T
$$
$$
+ S\mathrm{diag}(\mathbb{E}[a(X, u)])S^T. \tag{5}
$$

Above and in the sequel, time t is omitted from notation where no confusion may arise. If rates are affine in the state, i.e. $a(x, u) = W(u)x + w_0(u)$ for some

$W(u)$ and $w_0(u)$, these equations simplify to

$$\dot{\mu} = SW(u)\mu + Sw_0(u), \tag{6}$$

$$\dot{\Sigma} = SW(u)\Sigma + \Sigma W^T(u)S^T + S\mathrm{diag}\big(W(u)\mu + w_0(u)\big)S^T. \tag{7}$$

This system of differential equations is closed in the sense that it does not depend on unmodelled moments. If in addition W does not depend on u, then the system is linear in the input (and the initial conditions).

For the system (1)–(3), Eqs. (6)–(7) apply in the case of constant rates λ_+ and λ_-. In the general case of regulated switching rates $\lambda_+(X_?)$ and $\lambda_-(X_?)$, one may instead interpret (4)–(5) as the moment equations for the augmented state composed of X and $X_?$. Since the laws regulating $X_?$ are unspecified, the full system cannot be spelled out, but one may still work out the equations for the evolution of the moments of X_1, X_2 and X_3. Define

$$\big[z_{MP}^T \mid z_\times^T \mid z_F^T\big] = \big[\mu_M\, \mu_P\, \sigma_{MM}\, \sigma_{PP}\, \sigma_{MP} \mid \sigma_{MF}\, \sigma_{PF} \mid \mu_F\, \sigma_{FF}\big],$$

(vertical bars denoting vector blocks) where of course $\mu_\bullet$ and $\sigma_{\bullet\bullet}$ are the mean and covariance of the states corresponding to the species in subscript (identical subscripts denoting variance). From an engineering viewpoint, z_{MP} is the state of the dynamical sensor for the statistics of F, with sensor output given by the elements $[\mu_P\, \sigma_{PP}]^T$ of z_{MP}. Then one gets

$$\dot{z}_{MP} = A_{MP} \cdot z_{MP} + A_{MP,\times} \cdot z_\times + A_{MP,F} \cdot z_F, \tag{8}$$

$$\dot{z}_\times = A_\otimes \cdot z_\times + z_\otimes + A_{\times,F} \cdot z_F. \tag{9}$$

for matrices A_{MP}, $A_{MP,\times}$, $A_{MP,F}$, $A_\otimes$ and $A_{\times,F}$ depending solely on $\theta_{MP} = (k_M, d_M, k_P, d_P)$ (see Appendix A), i.e. the parameters of the sensing system. Note that, for every fixed t, $F(t)$ is a Bernoulli random variable. Then $\sigma_{FF}(t) = \mu_F(t)\big(1 - \mu_F(t)\big)$ for all t (as a consequence, (8)–(9) are somewhat redundant).

From (8)–(9) one observes that mean and variance of X_2, the observed elements of z_{MP}, are thus a dynamical transformation of those of F, i.e. z_F, plus a contribution from

$$z_\otimes = \mathrm{Cov}\left(\begin{bmatrix} X_1 \\ X_2 \end{bmatrix}, \begin{bmatrix} \lambda_+(X_?)(1 - X_3) \\ \lambda_-(X_?)X_3 \end{bmatrix}\right) \cdot \begin{bmatrix} 1 \\ -1 \end{bmatrix}.$$

As it will become clear, $z_\otimes$ implicitly brings about a contribution from the correlation structure of F (see later Remark 1).

2.3 Marginalization of Moments

From now on, abusing notation in favor of simplicity, we will refer to X_1, X_2 and X_3 by the symbols for the corresponding species, i.e. M, P and F, in the same order. Let f be any possible outcome of F, and let

$$\mu_P(t) = \mathbb{E}[P(t)], \qquad\qquad \mu_P^f(t) = \mathbb{E}[P(t)|F = f],$$

$$\mathscr{M}_P(t) = \mathbb{E}[P(t)^2], \qquad\qquad \mathscr{M}_P^f(t) = \mathbb{E}[P(t)^2|F = f],$$

where, unlike the approach in [7], conditioning is intended over the whole history of F. By marginalization,

$$\mu_P = \mathbb{E}\big[\mathbb{E}[P|F]\big] = \int \mu_P^f d\mathscr{P}_F(f), \quad \mathscr{M}_P = \mathbb{E}\big[\mathbb{E}[P^2|F]\big] = \int \mathscr{M}_Y^f d\mathscr{P}_F(f), \tag{10}$$

with $\mathscr{P}_F$ the probability distribution of F over all possible binary switching sequences. Let us now state the following assumption.

Assumption 1 (Granger causality [12]). *There is no feedback from M and P to F, i.e., at any time t, the future of F is conditionally independent on the past of M and P given the past of F.*

This captures the idea that species M and P do not participate in the regulation of the promoter [1,3], and corresponds well to all the reporter systems where reporter and regulatory proteins are physically different molecules. In the light of Assumption 1, the conditional moments μ_P^f and $\mathscr{M}_P^f$ are those of the reduced system (1)–(2) with f defining the state of species F at all times. Let

$$z_{MP}^f = [\mu_M^f \ \mu_P^f \ \sigma_{MM}^f \ \sigma_{PP}^f \ \sigma_{MP}^f]$$

be the vector of conditional moments of M and P. Working out the moment equations (6)–(7) for $X = [M \ P]^T$ and input $u = f$, one gets that

$$\dot{z}_{MP}^f = A_{MP} \cdot z_{MP}^f + (A_{MP,F})_1 \cdot f, \tag{11}$$

where $(A_{MP,F})_1$ denotes the first column of $A_{MP,F}$. Then μ_P^f and σ_{PP}^f follow from the solution of this system and $\mathscr{M}_P^f = \sigma_{PP}^f + (\mu_P^f)^2$, while marginalization (10) completes the computation of μ_P and $\mathscr{M}_P$. Note that, because of the relationship $\mathscr{M}_P = \sigma_{PP} + (\mu_P)^2$, we can equivalently consider (μ_P, σ_{PP}) or $(\mu_P, \mathscr{M}_P)$ to be the observed output quantities. We will often exploit this equivalence in the sequel without further notice.

Incidentally, notice that (11) represents a linear switching system with two alternating operational modes, $f = 0$ and $f = 1$.

3   The Fixed Rate Promoter Process

In order to investigate how statistics of F reflect into the observed profiles μ_P and $\mathscr{M}_P$, and how they may possibly be reconstructed from the output, we first focus on the fundamental case where switching rates λ_+ and λ_- are constant. Define $\alpha = \lambda_+ + \lambda_-$.

Proposition 1. *Mean $\mu_F(t) = \mathbb{E}[F(t)]$ and autocovariance function $\rho_F(t,s) = cov\big(F(t), F(s)\big)$ obey the equations*

$$\mu_F(t) = \mu_F(0)e^{-\alpha t} + \frac{\lambda_+}{\alpha}\left(1 - e^{-\alpha t}\right), \qquad\qquad t \geq 0, \quad (12)$$

$$\rho_F(t,\tau) = \left(\frac{\lambda_+}{\alpha} + \frac{(\alpha - \lambda_+)}{\alpha}e^{-\alpha(t-\tau)}\right) \cdot \mu_F(\tau) - \mu_F(t) \cdot \mu_F(\tau), \quad t \geq \tau. \quad (13)$$

In stationary conditions, with an abuse of notation for the arguments of ρ_F,

$$\mu_F = \frac{\lambda_+}{\alpha}, \qquad \rho_F(t-\tau) = \frac{\lambda_+(\alpha-\lambda_+)}{\alpha^2} \cdot e^{-\alpha(t-\tau)}. \qquad (14)$$

Incidentally, the autocovariance in (14) is the same as that of an Ornstein-Uhlenbeck process [15].

It can be appreciated that, in transient conditions, the mean profile μ_F contains all the information about the statistics of F. Indeed, in this simple case, rates λ_+ and λ_- (or equivalently α), together with the initial condition $\mu_F(0)$, fully determine the laws of F. In turn, these three quantities have distinct effects on μ_F, i.e. they are distinguishable from a transient mean profile. In [2], it was shown that these and other model parameters, notably θ_{MP}, are also jointly distinguishable from the measured profiles μ_P and $\mathcal{M}_P$. The result is based on the specialization of (8)–(9) for the case of the fixed rate process, given by [2]

$$\begin{bmatrix} \dot{z}_{MP} \\ \dot{z}_\times \\ \dot{z}_F \end{bmatrix} = \begin{bmatrix} A_{MP} & A_{MP,\times} & A_{MP,F} \\ 0 & A_\times & A_{\times,F} \\ 0 & 0 & A_F \end{bmatrix} \cdot \begin{bmatrix} z_{MP} \\ z_\times \\ z_F \end{bmatrix} + \begin{bmatrix} 0 \\ 0 \\ u_F \end{bmatrix}, \qquad (15)$$

where $A_\times = -\alpha I + A_\otimes$, $u_F = [\lambda_+ \ \lambda_+]^T$ and A_F depends only on λ_+ and α as detailed in Appendix A. For known parameters θ_{MP}, we may easily show that λ_-, λ_+ and $\mu_F(0)$ are also distinguishable from the sole mean μ_P. For simplicity, we consider the case where M and P are identically 0 at time 0. By inspection of (15),

$$\begin{aligned} \dot{\mu}_F &= -\alpha\mu_F + \lambda_+, \\ \dot{\mu}_M &= -d_M\mu_M + k_M\mu_F, \qquad (16) \\ \dot{\mu}_P &= -d_P\mu_P + k_P\mu_M \end{aligned}$$

(the expression of $\dot{\mu}_F$ above coincides with the differential form of (12)). Thus, in terms of Laplace transform,

$$\mu_P(s) = \frac{\lambda_+ k_M k_P}{s(\alpha+s)(d_M+s)(d_P+s)} + \frac{\mu_F(0)k_M k_P}{(\alpha+s)(d_M+s)(d_P+s)},$$

and one may apply the method of [2] (also reported in Appendix B) to prove sensitivity of this solution (equivalently, the solution over time) to any change in the three unknown parameters, almost everywhere in the space of the remaining parameters. In practical terms, parameter values can be reconstructed either from μ_F as obtained by deconvolution from μ_P, or by direct fit of (16) to an observed μ_P profile.

Now assume that F has reached stationarity. In this case, all relevant statistics of F are determined by λ_+ and λ_-. However, from Proposition 1, mean μ_F only conveys information about the ratio λ_+/α, and, because $\sigma_F^2 = \mu_F(1-\mu_F)$ at any point in time, no more information is contained in the variance. Specific contributions of the two parameters can instead be traced in the autocovariance function ρ_F. Indeed, multiplicative factor $\lambda_+(\alpha-\lambda_+)/\alpha^2$ and decay rate α have distinguishable effects on ρ_F (different choices of the two lead to different profiles $\rho_F(\cdot)$) and uniquely determine λ_+ and λ_-. The question arises whether ρ_F is observable from the measured profiles μ_P and $\mathcal{M}_P$ (i.e. whether λ_+ and λ_- are also distinguishable from the experimental output). In this section we provide a positive answer in terms of identifiability of λ_+ and λ_-, i.e. for processes F with fixed rates. A more general answer will be provided in the next section.

From Proposition 1, stationary conditions are achieved when μ_F is in steady state (i.e. when the factors of $\rho_F(t, \tau)$ involving μ_F no longer depend on τ). It then suffices to check identifiability of λ_+ and λ_- from the solution of (15) with stationary initial conditions $\mu_F(0) = \lambda_+/\alpha$ and $\sigma_F^2(0) = \lambda_+/\alpha(1 - \lambda_+/\alpha)$. Using again the method of [2], one computes the Laplace response function of this system. The resulting equations are lengthy and not reported here. Then, it can be checked that the Laplace sensitivity condition also reported in Appendix B is verified, i.e. the time profiles of μ_P and $\mathcal{M}_P$ are sensitive to all possible changes of λ_+ and α, almost everywhere in the space of the parameters θ_{MP}.

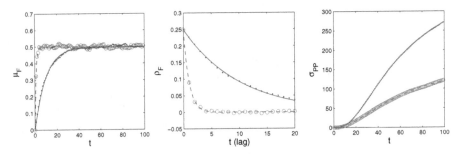

Fig. 1. Statistics for F (dashed lines and circles) and F' (solid lines and dots). Lines visualize analytic solutions, markers are for empirical statistics from Gillespie simulations. Gillespie simulations are performed using Stochkit [17] for the generation of 10^4 sample paths (i.e. simulated cells). Numerical calculations are performed in Matlab.

Example 1. Refer to Fig. 1. Statistics for two fixed-rate promoter activity processes, F and F', are considered. F has $\lambda_+ = \lambda_- = 0.05$, while F' has the faster switching dynamics $\lambda_+ = \lambda_- = 0.5$. Starting from the non-stationary conditions $F = F' = 0$ at time 0, means μ_F and $\mu_{F'}$ converge both at 0.5 at different rates (Fig. 1, left), thus resulting into different output profiles μ_P (not shown). In other words, the two processes are distinguishable from the mean. In stationary conditions, instead, the means for F and F' are the same. Yet the stationary autocovariance functions ρ_F and $\rho_{F'}$ differ in the two cases (Fig. 1, center). This results in different observed profiles of σ_{PP} (Fig. 1, right). In other words, in stationary conditions, F and F' are distinguishable from the output variance.

Remark 1. Equations (15) are obtained from (8)–(9) by developing the expression of $z_\otimes$. This results in expressions depending on matrices $A_\times$ and A_F, which bring in the role of α, the decay rate of ρ_F, into the propagation of second-order moments from z_F to z_{MP}. This fact is indeed in agreement with the discussion of $z_\otimes$ at the end of Sect. 2.2.

To summarize, we have shown that constant switching rates, whence all statistics, of a promoter activity process F can be reconstructed from the output mean if F is not in stationary conditions. In stationary conditions, the promoter statistics cannot be determined from the output mean, but rather from the output variance since this reflects differences in the autocovariance function of F. Analytic expressions and a case study have been developed to support our arguments.

4   General Promoter Switching Processes

We now wish to study how first- and second-order moments of switching process F reflect into outputs μ_P and $\mathscr{M}_P$, and how to possibly reconstruct the former from the latter, without a priori knowledge on F. In particular, we do not assume that switching rates λ_+ and λ_- are fixed. We only assume that F has continuous (mean and) autocovariance $\rho_F(t, s)$, and that Assumption 1 holds. For simplicity, we focus on the case where $z_{MP}(0)$ is null (M and P equal to zero at time zero).

From Eq. (11), for some final time $T > 0$, the conditional moments $\mu_P^f(t)$ and $\sigma_{PP}^f(t)$ over $[0, T)$ are the output of a linear dynamical system with (zero initial conditions and) input f. We may then introduce linear operators, L_1 and L_2, and abstract the transformation from function f to μ_P^f and σ_{PP}^f as $\mu_P^f = L_1 f$ and $\sigma_{PP}^f = L_2 f$. When necessary, for any $t \in [0, T)$, we will write $\mu_P^f(t) = (L_1 f)(t)$ as $L_1^t f$ and $\sigma_{PP}^f(t) = (L_2 f)(t)$ as $L_2^t f$. Of course, for $k = 1$ and $k = 2$,

$$L_k^t f = \int_0^t d\tau\, \ell_k(t, \tau) f(\tau), \qquad \ell_k(t, \tau) = C_k e^{A_{MP}(t-\tau)} (A_{MP,F})_1,$$

with $C_1 = \begin{bmatrix} 0 & 1 & 0 & 0 & 0 \end{bmatrix}$ (mean readout) and $C_2 = \begin{bmatrix} 0 & 0 & 0 & 1 & 0 \end{bmatrix}$ (variance readout).

4.1   Observability and Reconstruction of the Process Mean

From the first equality in (10), one has that

$$\mu_P = \int (L_1 f) d\mathscr{P}_F(f) = L_1 \left(\int f d\mathscr{P}_F(f) \right) = L_1 \mu_F.$$

Not surprisingly at this point, μ_P thus follows from the linear dynamical transformation of μ_F already found in (8). Observability of μ_F from μ_P essentially depends on the spectrum of L_1. Since

$$\mu_P(s) = \frac{k_M k_P}{(d_M + s)(d_P + s)} \mu_F(s),$$

for strictly positive parameters θ_{MP}, the transformation is invertible over the whole spectrum, i.e. μ_F can be perfectly reconstructed from μ_P. In practice, this amounts to a deconvolution problem of rather easy solution [2].

4.2 Observability and Reconstruction of the Process Covariance

We begin with the following result.

Proposition 2. *For any time $t \in [0, T)$, it holds that*

$$\mathscr{M}_P(t) = L_2^t \mu_F + \mathbb{E}[(L_1^t F)^2]. \tag{17}$$

Clearly the autocovariance function of F plays a role in the term $\mathbb{E}[(L_1^t F)^2]$. To study this term further, consider the Karhunen-Loève decomposition [15] of process F, given by

$$F - \mu_F = \sum_{i=1}^{\infty} a_i \phi_i,$$

where the ϕ_i are the mutually orthogonal, unit norm eigenfunctions of the operator $K : \phi \mapsto \int d\tau \rho_F(\cdot, \tau)\phi(\tau)$, i.e. $K\phi_i = \sigma_i^2 \phi_i$, and the a_i are mutually uncorrelated, zero-mean random variables with variance equal to the eigenvalues σ_i^2 (function norm is in L^2 and the decomposition holds in the mean-square sense). Then

$$\rho_F(t, \tau) = \sum_{i=1}^{\infty} \sigma_i^2 \phi_i(t)\phi_i(\tau), \qquad \sigma_{FF}(t) = \sum_{i=1}^{\infty} \sigma_i^2 \phi_i^2(t).$$

Proposition 3. *It holds that* $\mathbb{E}[(L_1^t F)^2] = (L_1^t \mu_F)^2 + \sum_{i=1}^{\infty} \sigma_i^2 (L_1^t \phi_i)^2$.

In sums, from Propositions 2–3 and using the fact that $(\mu_P)^2 = (L_1 \mu_F)^2$,

$$\sigma_{PP}(t) = \mathscr{M}_P(t) - \mu_P^2(t) = L_2^t \mu_F + \sum_{i=1}^{\infty} \sigma_i^2 (L_1^t \phi_i)^2. \tag{18}$$

Comparing the expressions of σ_{PP} and σ_{FF} one notices that, besides term $L_2^t \mu_F$, the functions composing F and characterizing its autocovariance structure are transformed by L_1^t into contributions that make up the variance of P at time t. Were L_1^t an evaluation operator, i.e. $L_1^t \phi_i = \phi_i(t)$, then $\sigma_{PP}(t)$ would degenerate to $L_2^t \mu_F + \sigma_{FF}(t)$, i.e. information about the autocovariance structure of F would be lost. For every t, it is the integral nature of L_1^t that channels information about the whole $\rho_F(\cdot, \cdot)$ into $\sigma_{PP}(t)$. Another viewpoint on this is given in what follows.

Equation (18) explains the nature of the information transfer from ρ_F to σ_{PP}. For reconstruction purposes, however, we seek a more explicit relationship between σ_{PP} and ρ_F. The following result relies on the convolutional form of L_1.

Proposition 4. *It holds that*

$$\sigma_{PP}(t) = L_2^t \mu_F + \iint d\tau \, dv \, \ell_1(t, \tau)\ell_1(t, v)\rho_F(\tau, v). \tag{19}$$

Hence ρ_F undergoes itself a linear transformation H defined by

$$H^t\rho = (H\rho)(t) = \int_0^t d\tau \int_0^t dv\, \ell_1(t,\tau)\ell_1(t,v)\rho(\tau,v).$$

In particular, suppose that F is stationary. Then, by a change of variables,

$$H^t\rho_F = \int_0^t d\tau \int_0^t dv\, \ell_1(t,\tau)\ell_1(t,v)\rho_F(\tau - v) = \int_{-t}^t d\delta\, h(t,\delta)\rho_F(\delta)$$

with

$$h(t,\delta) = \int_{\max\{-\delta,0\}}^{\min\{t,t-\delta\}} dv\, \ell_1(t,v+\delta)\ell_1(t,v).$$

In the light of these results, the problem of the observability of ρ_F, or better the joint observability of ρ_F and μ_F from μ_P and σ_{PP}, is thus equivalent to that of the invertibility of the linear operator

$$(\mu_F, \rho_F) \mapsto (L_1\mu_F, L_2\mu_F + H\rho_F) \tag{20}$$

(with relevant simplifications if stationarity of F is hypothesized). We note that, besides term $L_2\mu_F$, the relationship between ρ_F and σ_{PP} is analogous to that pertaining linear transformations of second-order processes. In particular, using the fact that $\ell_1(t,\cdot)$ is the impulse response of a time-invariant dynamical system, the second term of (19) can be seen as the autocovariance of the output of a linear filter with response ℓ_1 fed with an input process with autocovariance ρ_F. It is then natural to frame observability analysis of ρ_F in the context of spectral analysis [11,12]. This analysis is left for future work. Here we limit ourselves to the discussion of an illustrative example.

Example 2. Refer to Fig. 2. We consider a promoter activity process F with randomly regulated rates, and compare its statistics with those of relevant fixed-rate processes F' and F''. All processes are analyzed in stationary regime and have rate λ_- identically set to 0.5, i.e. their definition only differs in the activation rate. The activation rate of F is $\lambda_+(R) = 1 \cdot R$. Regulator R is another random binary process with switch-off rate equal to 0.1 and switch-on rate, equal to 0.2217, chosen so as to guarantee that the stationary mean of F is $\mu_F = 0.5$. Process F' is defined as in Example 1, i.e. it has $\lambda_+ = 0.5$, again resulting in $\mu_{F'} = 0.5$. Finally, process F'' has activation rate $\lambda_+ = \mathbb{E}[\lambda_+(R)] = 0.6892$, i.e. a switch-on rate equivalent on average to that of F. This results in a different mean, $\mu_{F''} = 0.5795$.

The autocovariance function of F (Fig. 2, left) is markedly different from those of F' and F'', which are similar. Because $\mu_F = \mu_{F'} \neq \mu_{F''}$, F can be distinguished from F'', but not from F', from the output mean μ_P (Fig. 2, center). However, because of the different autocovariance function, F can be distinguished from F' from the output variance σ_{PP}. Interestingly, the output variance profiles for F and F'' are quite similar, a sign that the differences

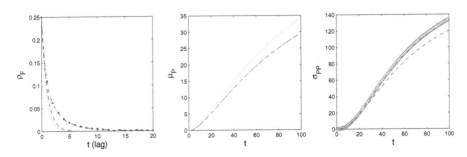

Fig. 2. Statistics for a random-rate promoter process F (dash-dotted lines) and relevant fixed-rate promoter processes F' (same as in Fig. 1, dashed lines) and F'' (dotted lines). Left: autocovariance functions ρ_F (dots: estimates from Gillespie simulations; line: interpolation), $\rho_{F'}$ (from (14)) and $\rho_{F''}$ (from (14)); Center: Observed output mean μ_P for F (Gillespie simulation), F' (solution of (15)) and F'' (solution of (15)) – curves for F and F' are superimposed; Right: The observed output variance of P for F (diamonds: numerical computation of (19), based on the profile of ρ_F from Gillespie simulations; line: estimate from Gillespie simulation – diamonds and line are superimposed), F' (solution of (15)) and F'' (solution of (15)) – curves for F and F'' are superimposed. Gillespie simulations are performed using Stochkit [17] for the generation of 10^5 sample paths (i.e. simulated cells). Numerical calculations are performed in Matlab.

between F and F'' in mean and autocovariance compensate each other in this case. This is possible since output variance depends not only on the autocovariance but also on the mean of the promoter activity process.

Finally, in the light of the linearity of (20), joint estimation of μ_F and ρ_F from possibly noisy and sampled measurements of μ_P and σ_{PP} can be seen as a linear inversion problem. Regularized solutions for both the stationary and the nonstationary case may be developed in accordance with the vast literature on the subject (see e.g. [4] and references therein). Note that, because μ_F can be reconstructed from the sole mean μ_P, the problem may also be reduced to that of estimating ρ_F from $\sigma_{PP} - L_2^t \mu_F$ via (regularized) inversion of H.

In summary, we have analyzed the relationship between second-order promoter activity statistics and mean and variance profiles of the reporter protein P in the case of promoter processes with randomly regulated rates. In particular, we have developed explicit relationships between the autocovariance function of F and the readout variance profile of P, showing that this integral relationship is essentially linear. By this we provided the basis for a full spectral analysis of observability of ρ_F and its linear reconstruction from reporter protein mean and variance statistics. We also illustrated the relevance of our results by investigating the distinguishability of a random-rate and relevant fixed-rate processes on an example.

5   Conclusions

We have studied the relationships between second-order statistics of random promoter activity and the mean and variance profiles of gene expression reporter proteins typically observed in biological experiments. For both fixed and randomly regulated (thus also possibly time-varying) switching rates, we developed explicit mathematical formulas showing that these relationships are linear, and provided first results about the observability of the promoter process statistics from gene reporter data. Based on analytic considerations as well as on example case studies, we showed when and how analysis of second-order moments allows for discrimination of different promoter activation statistics.

This work provides the basis for an extensive observability analysis of promoter processes from gene reporter data at a single-cell level, and the development of promoter statistics reconstruction algorithms that are fully nonparametric, i.e. independent of a priori knowledge about the promoter activity laws. Our results show that both observability and estimation can be framed in the well-studied context of linear operators. Subsequent research work will henceforth focus on the application of the relevant spectral analysis and regularized linear inversion techniques. On these bases, we will then address a key challenge of this research effort, namely the identification and discrimination among alternative promoter activity regulatory mechanisms on the basis of the reconstructed promoter activation statistics and data from candidate regulators.

A   Definitions and Proofs

Matrix definitions. A_{MP} $A_{MP,\times}$ $A_{MP,F}$ are given by

$$
\begin{bmatrix}
-d_M & 0 & 0 & 0 & 0 \\
k_P & -d_P & 0 & 0 & 0 \\
d_M & 0 & -2d_M & 0 & 0 \\
k_P & d_P & 0 & -2d_P & 2k_P \\
0 & 0 & k_P & 0 & -d_M - d_P
\end{bmatrix},
\begin{bmatrix}
0 & 0 \\
0 & 0 \\
2k_M & 0 \\
0 & 0 \\
0 & k_M
\end{bmatrix},
\begin{bmatrix}
k_M & 0 \\
0 & 0 \\
k_M & 0 \\
0 & 0 \\
0 & 0
\end{bmatrix},
$$

in the same order, while

$$
A_\otimes = \begin{bmatrix} -d_M & 0 \\ k_P & -d_P \end{bmatrix}, \qquad
A_{\times,F} = \begin{bmatrix} 0 & k_M \\ 0 & 0 \end{bmatrix}, \qquad
A_F = \begin{bmatrix} -\alpha & 0 \\ \alpha - 2\lambda_+ & -2\alpha \end{bmatrix}.
$$

Proof of Proposition 1. Process F is a homogeneous continuous-time binary Markov chain. Letting $p(t) = \begin{bmatrix} \text{Prob}\{F(t) = 0\} & \text{Prob}\{F(t) = 1\} \end{bmatrix}^T$, for any t and τ it holds that

$$
p(t) = e^{Q(t-\tau)}p(\tau), \qquad\qquad Q = \begin{bmatrix} -\lambda_+ & \lambda_- \\ \lambda_+ & -\lambda_- \end{bmatrix}.
$$

Mean $\mu_F = \text{Prob}\{F(t) = 1\}$. Using the fact that $\dot{p} = Qp$, the differential equation for μ_F, the second element of p, is $\dot{\mu}_F = \lambda_+(1 - \mu_F) -$

$\lambda_- \mu_F = -\alpha \mu_F + \lambda_+$. The solution of this equation relative to $\mu_F(0)$ yields the expression in the statement. Covariance $\rho_F(t,\tau) = \text{Prob}\{F(t) = 1, F(\tau) = 1\} - \mu_F(t)\mu_F(\tau)$. By Bayes'law, $\text{Prob}\{F(t) = 1, F(\tau) = 1\} = \text{Prob}\{F(t) = 1|F(\tau) = 1\} \cdot \text{Prob}\{F(\tau) = 1\}$. Second factor is equal to $\mu_F(\tau)$, while the first factor is given by the entry of row 2 and column 1 of $e^{Q(t-\tau)}$. Computing the matrix exponential thus yields the result. Stationary versions of μ_F and ρ_F are found simply by taking the limit of $\mu_F(t)$ as $t \to +\infty$ and replacing the result for $\mu_F(\tau)$ and $\mu_F(t)$ in the expression of $\rho_F(t,\tau)$.

Proof of Proposition 2. Starting from the second relation in (10),

$$
\begin{aligned}
\mathcal{M}_P(t) &= \mathbb{E}\big[\mathbb{E}[P^2|F]\big] = \mathbb{E}\Big[\mathbb{E}\big[\big((P - \mathbb{E}[P|F]) + \mathbb{E}[P|F]\big)^2|F\big]\Big] \\
&= \mathbb{E}\Big[\mathbb{E}\big[(P - \mathbb{E}[P|F])^2|F\big]\Big] + \mathbb{E}\Big[\mathbb{E}\big[\mathbb{E}[P|F]^2|F\big]\Big] \\
&\quad + 2 \cdot \mathbb{E}\Big[\mathbb{E}\big[(P - \mathbb{E}[P|F]) \cdot \mathbb{E}[P|F]|F\big]\Big] \\
&= \mathbb{E}\Big[\mathbb{E}\big[(P - \mathbb{E}[P|F])^2|F\big]\Big] + \mathbb{E}\Big[\mathbb{E}[P|F]^2\Big] \\
&\quad + 2 \cdot \mathbb{E}\Big[\mathbb{E}\big[P - \mathbb{E}[P|F]|F\big] \cdot \mathbb{E}[P|F]\Big],
\end{aligned}
$$

where the last row vanishes since $\mathbb{E}\big[P - \mathbb{E}[P|F]|F\big] = 0$. Then, using the definitions of μ_P^F and σ_{PP}^F, the chain of equalities continues with

$$
= \mathbb{E}\big[\sigma_{PP}^F(t)\big] + \mathbb{E}\big[(\mu_P^F(t))^2\big] = \mathbb{E}\big[L_2^t F\big] + \mathbb{E}\big[(L_1^t F)^2\big] = L_2^t \mu_F + \mathbb{E}[(L_1^t F)^2].
$$

Proof of Proposition 3. The following chain of inequalities hold:

$$
\begin{aligned}
\mathbb{E}[(L_1^t F)^2] &= \mathbb{E}\left[\left(L_1^t\left(\sum_i a_i \phi_i\right)\right)^2\right] = \mathbb{E}\left[\sum_{i,j} L_1^t(a_i \phi_i) L_1^t(a_j \phi_j)\right] \\
&= \sum_{i,j} \mathbb{E}[a_i a_j](L_1^t \phi_i)(L_1^t \phi_j) = \sum_i \sigma_i^2 (L_1^t \phi_i)^2,
\end{aligned} \tag{21}
$$

where the latter equality follows from the mutual uncorrelation of the a_i.

Proof of Proposition 4. Expanding the last term of (17) one gets

$$
\begin{aligned}
\mathbb{E}[(L_1^t F)^2] &= \int d\mathscr{P}_F(f)(L_1^t f)^2 \\
&= \int d\mathscr{P}_F(f)\left(\int_0^t d\tau\, \ell_1(t,\tau) f(\tau)\right)\left(\int_0^t dv\, \ell_1(t,v) f(v)\right) \\
&= \int_0^t d\tau \int_0^t dv\, \ell_1(t,\tau)\ell_1(t,v)\left(\int d\mathscr{P}_F(f) f(\tau) f(v)\right) \\
&= \int_0^t d\tau \int_0^t dv\, \ell_1(t,\tau)\ell_1(t,v)\big(\rho_F(\tau,v) + \mu_F(\tau)\mu_F(v)\big)
\end{aligned}
$$

where the last integrand is of course the autocorrelation of F at τ and v. Therefore

$$\sigma_{PP}(t) = \mathcal{M}_P(t) - \mu_P^2(t)$$

$$= L_2^t \mu_F + \int_0^t d\tau \int_0^t dv \, \ell_1(t,\tau)\ell_1(t,v)\big(\rho_F(\tau,v) + \mu_F(\tau)\mu_F(v)\big)$$

$$- \left(\int_0^t d\tau \, \ell_1(t,\tau)\mu_F(\tau)\right)\left(\int_0^t dv \, \ell_1(t,v)\mu_F(v)\right),$$

and the result follows by collecting integrals and simplifying.

B   Laplace Sensitivity Method for the Analysis of Parameter Identifiability

This section reports the identifiability analysis method of [2]. Let $\mathcal{Y}_\theta(t)$ be a vector function of $t \in \mathbb{R}$ depending on parameters θ. Typically $\mathcal{Y}_\theta(\cdot)$ is an observed response of a dynamical system defined in terms of θ.

Definition 1. *The parametric family (of functions) $\{\mathcal{Y}_\theta : \theta \in \Theta\}$, with $\Theta \subseteq \mathbb{R}^N$, $N \in \mathbb{N}$, is*

(a) locally identifiable at θ^* if a neighborhood $B_{\theta^*} \subseteq \Theta$ of θ^* exists such that the implication holds $\forall \theta \in B_{\theta^*}$;

(b) locally identifiable if (a) holds for almost every (a.e.) $\theta^* \in \Theta$.

For any given θ let $Y(s,\theta)$ be the Laplace transform of $\mathcal{Y}_\theta(\cdot)$. Let $\nabla Y(s,\theta) = \frac{\partial Y}{\partial \theta}(s,\theta) = \left[\frac{\partial Y}{\partial \theta_1} \cdots \frac{\partial Y}{\partial \theta_N}\right](s,\theta)$.

Proposition 5. *If, for some $L \in \mathbb{N}$, a set of points $\mathscr{S}_L = \{s_1, \dots, s_L\} \subseteq \mathbb{R}$ (or $\mathbb{C}$) exists such that the matrix*

$$\Delta(\mathscr{S}_L, \theta^*) = \left[\nabla Y(s_1, \theta^*)^T \cdots \nabla Y(s_L, \theta^*)^T\right]^T$$

has full column rank, then $\{\mathcal{Y}_\theta : \theta \in \Theta\}$ is locally identifiable at θ^* (in the sense of Definition 1(a)).

Now assume that the elements of $Y(s,\theta)$ are ratios of polynomials in the entries of θ.

Corollary 1. *If, for a given set of points $\mathscr{S}_L$ and a given θ^*, matrix $\Delta(\mathscr{S}_L, \theta^*)$ is full column rank, then $\{\mathcal{Y}_\theta : \theta \in \Theta\}$ is locally identifiable (a.e. in the sense of Definition 1(b)).*

In the present paper, the Laplace transforms that are used to discuss identifiability belong to this last class (see [2]), whence Corollary 1 applies. In practice, these conditions can be easily checked by the use of the Matlab Symbolic Math Toolbox and evaluation of the rank conditions based on a finite set of heuristically chosen points $\mathscr{S}_L$ (see again [2]).

References

1. Bowsher, C.G., Voliotis, M., Swain, P.S.: The fidelity of dynamic signaling by noisy biomolecular networks. PLoS Comput. Biol. **9**(3), e1002965 (2013)
2. Cinquemani, E.: Reconstruction of promoter activity statistics from reporter protein population snapshot data. In: 2015 54th IEEE Conference on Decision and Control (CDC), pp. 1471–1476, December 2015
3. Cinquemani, E.: Reconstructing statistics of promoter switching from reporter protein population snapshot data. In: Abate, A., et al. (eds.) HSB 2015. LNCS, vol. 9271, pp. 3–19. Springer, Heidelberg (2015). doi:10.1007/978-3-319-26916-0_1
4. De Nicolao, G., Sparacino, G., Cobelli, C.: Nonparametric input estimation in physiological systems: problems, methods, and case studies. Automatica **33**(5), 851–870 (1997)
5. Friedman, N., Cai, L., Xie, X.S.: Linking stochastic dynamics to population distribution: an analytical framework of gene expression. Phys. Rev. Lett. **97**, 168302 (2006)
6. Hasenauer, J., Waldherr, S., Doszczak, M., Radde, N., Scheurich, P., Allgower, F.: Identification of models of heterogeneous cell populations from population snapshot data. BMC Bioinform. **12**(1), 125 (2011)
7. Hasenauer, J., Wolf, V., Kazeroonian, A., Theis, F.J.: Method of conditional moments (MCM) for the chemical master equation. J. Math. Biol. **69**(3), 687–735 (2014)
8. Hespanha, J.: Modelling and analysis of stochastic hybrid systems. IEE Proc. Control Theor. Appl. **153**(5), 520–535 (2006)
9. de Jong, H., Ranquet, C., Ropers, D., Pinel, C., Geiselmann, J.: Experimental and computational validation of models of fluorescent and luminescent reporter genes in bacteria. BMC Syst. Biol. **4**(1), 55 (2010)
10. Kaern, M., Elston, T.C., Blake, W.J., Collins, J.J.: Stochasticity in gene expression: from theories to phenotypes. Nat. Rev. Gen. **6**, 451–464 (2005)
11. Koopmans, L.H.: The Spectral Analysis of Time Series. Probability and Mathematical Statistics. Academic Press, San Diego (1995)
12. Lindquist, A., Picci, G.: Linear Stochastic Systems - A Geometric Approach to Modeling, Estimation and Identification. Springer, Heidelberg (2015)
13. Munsky, B., Trinh, B., Khammash, M.: Listening to the noise: Random fluctuations reveal gene network parameters. Mol. Syst. Biol. **5** (2009). Article ID 318
14. Neuert, G., Munsky, B., Tan, R., Teytelman, L., Khammash, M., van Oudenaarden, A.: Systematic identification of signal-activated stochastic gene regulation. Science **339**(6119), 584–587 (2013)
15. Papoulis, A.: Probability, Random Variables, and Stochastic Processes. McGraw-Hill Series in Electrical Engineering. McGraw-Hill, New York (1991)
16. Paulsson, J.: Models of stochastic gene expression. Phys. Life Rev. **2**(2), 157–175 (2005)
17. Sanft, K.R., Wu, S., Roh, M., Fu, J., Lim, R.K., Petzold, L.R.: Stochkit2: Software for discrete stochastic simulation of biochemical systems with events. Bioinformatics **27**(17), 2457–2458 (2011)
18. Stefan, D., Pinel, C., Pinhal, S., Cinquemani, E., Geiselmann, J., de Jong, H.: Inference of quantitative models of bacterial promoters from time-series reporter gene data. PLoS Comput. Biol. **11**(1), e1004028 (2015)
19. Zechner, C., Ruess, J., Krenn, P., Pelet, S., Peter, M., Lygeros, J., Koeppl, H.: Moment-based inference predicts bimodality in transient gene expression. PNAS **21**(109), 8340–8345 (2012)

20. Zechner, C., Unger, M., Pelet, S., Peter, M., Koeppl, H.: Scalable inference of heterogeneous reaction kinetics from pooled single-cell recordings. Nat. Methods **11**, 197–202 (2014)
21. Zulkower, V., Page, M., Ropers, D., Geiselmann, J., de Jong, H.: Robust reconstruction of gene expression profiles from reporter gene data using linear inversion. Bioinformatics **31**(12), i71–i79 (2015)

Logic-Based Multi-objective Design of Chemical Reaction Networks

Luca Bortolussi[1,2,3], Alberto Policriti[4,6], and Simone Silvetti[4,5(✉)]

[1] DMG, University of Trieste, Trieste, Italy
luca@dmi.units.it
[2] Modelling and Simulation Group, Saarland University, Saarbrücken, Germany
[3] CNR-ISTI, Pisa, Italy
[4] Dima, University of Udine, Udine, Italy
alberto.policriti@uniud.it
[5] Esteco SpA, Trieste, Italy
silvetti@esteco.com
[6] Istituto di Genomica Applicata, Udine, Italy

Abstract. The design of genetic or protein networks that satisfy a given set of behavioural specifications is one of the main challenges of synthetic biology. Model-based design is a natural choice in this respect. Here we consider the problem of tuning parameters of a stochastic model to force one or more behavioural goals to hold. In particular, we consider several objectives specified by signal temporal logic formulae, and we look for a parameter set making their satisfaction probability as large as possible. This formalisation results in a multi-objective optimisation problem, which we solve by considering an optimisation scheme combining satisfaction probability and average robustness of STL properties, leveraging state of the art multi-objective optimisation routines.

Keywords: System design · Robustness · Multi-objective optimization · Temporal logic

1 Introduction

In the age of synthetic biology, we have the ability of building in vivo novel genetic pathways, and to engineer cellular behaviours. Due to the cost and complexity of biological systems, the standard is to rely on model-based design. As typical in Computer-Aided Engineering, this process consists in building a model of the system of interest, in formalizing objectives to be achieved, in tuning controllable parameters to satisfy the desired goals as robustly as possible. We refer to this last step as system design.

In systems and synthetic biology, one typically considers stochastic models described as Chemical Reaction Networks (CRN, [18]), resulting in Continuous Time Markov Processes that account for the intrinsic noise in stochastic

LB acknowledges partial support from EU-FET project QUANTICOL (nr. 600708) and from FRA-UniTS.

© Springer International Publishing AG 2016
E. Cinquemani and A. Donzé (Eds.): HSB 2016, LNBI 9957, pp. 164–178, 2016.
DOI: 10.1007/978-3-319-47151-8_11

systems [23]. CRN can be simulated [7,18], or approximated by mean-field or moment closure techniques [6]. More refined analysis of these models can be obtained by leveraging tools for formal verification of stochastic systems [1]. This process consists in expressing properties by means of a formal language like temporal logic, and by using numerical or statistical model checking algorithms to estimate the satisfaction probabilities of such properties, for a given model [3,7]. A natural intersection between verification and system design is to use temporal logic properties to specify the desired behaviours, tuning model parameters to optimise their satisfaction probability [4]. A successful application of this idea [4] exploits a quantitative semantics of signal temporal logic [14], which return a measure of robustness of the (dis)satisfaction of a formula.

In this article we will push forward the use of logical methods for systems design, encapsulating them into a multi-objective approach. The idea is to discover parameter values of a stochastic model of a CRN leading to the best trade off in the maximization of the probability satisfaction of different formulae, encoding contrasting objectives to be achieved. We will pursue this goal by combining both the satisfaction probability and the average robustness score of a formula and leveraging state of the art multi-objective routines, in order to achieve a more effective exploration of the parameter space. We will further discuss the results of [4] in terms of optimisation of single objectives, focussing on a potential pitfall in the use of the robustness score, related to a difference in the scale of atomic propositions, and showing the benefits of multi-objective optimisation.

We will start the paper by recalling the needed basic concepts (Sect. 2), and then introducing two case studies (a SIRS Epidemic Model and a Genetic Toggle Switch) that we will use to exemplify our approach (Sect. 3). In Sect. 4 we give the definition of the multi-objective problem we want to solve and three different strategies we have tried. Results on the case studies are discussed in Sect. 5. Section 6, instead, is devoted to the comparison of single and multi-objective optimisation in terms of the quantitative semantics. Conclusions and future work are discussed in Sect. 7.

Related Work. Multi-objective optimization techniques have been widely studied over the last fifteen years due to their increasing importance in many academic research fields (economics, operational research, stochastic control theory, etc.) as well in industrial research (see [10]). Several solution methods exist (e.g. evolutionary algorithms [11,24], gradient optimizer with restart, single weighted sum), mostly based on heuristic search. The multi-objective approaches to model checking verification have been described in [9,15,16]. The most common technique consists in transforming the original problem in a linear programming problem. The system design of stochastic models by using the robustness of temporal properties is a rather new field of research [4] and in particular the multi-objective approaches in this case have not been yet explored.

2 Background

In this section we introduce the relevant background material: Parametric Chemical Reaction Networks, Signal Temporal Logic, and multi-objective optimization.

2.1 Parametric Chemical Reaction Networks

Chemical Reaction Networks [18] are a standard model of population processes, known in literature also as Population Continuous Time Markov Chains [6] or Markov Population Models [19]. We consider a slight variant, explicitly rendering kinetic parameters.

Definition 1. *A Parametric Chemical Reaction Network (PCRN) $\mathcal{M}$ is a tuple* $(\mathcal{S}, \mathbf{X}, D, \mathbf{x_0}, \mathcal{R}, \Theta)$

- *$\mathcal{S} = \{s_1, \ldots, s_n\}$ is the set of species;*
- *$\mathbf{X} = (X_1, \ldots, X_n)$ is the vector of variables counting the amount of each species, with values $\mathbf{X} \in D$, where $D = \mathbb{N}^n$ is state space;*
- *$\mathbf{x_0} \in D$ is the initial state;*
- *$\mathcal{R} = \{r_1, \ldots, r_m\}$ is the set of chemical reactions, each of the form $r_j = (\mathbf{v_j}, \alpha_j)$, with $\mathbf{v_j}$ the stoichiometry or update vector and $\alpha_j = \alpha_j(\mathbf{X}, \theta)$ the propensity or rate function. Each reaction can be represented as*

$$r_j : u_{j,1}s_1 + \ldots + u_{j,n}s_n \xrightarrow{\alpha_j} w_{j,1}s_1 + \ldots + w_{j,n}s_n,$$

where $u_{j,i}$ ($w_{j,i}$) is the amount of elements of species s_i consumed (produced) by reaction r_j. We let $\boldsymbol{u_j} = (u_{j,1}, \ldots, u_{j,n})$ (and similarly $\boldsymbol{w_j}$) and define $\boldsymbol{v_j} = \boldsymbol{w_j} - \boldsymbol{u_j}$.
- *$\theta = (\theta_1, \ldots, \theta_k)$ is the vector of (kinetic) parameters, taking values in a compact hyperrectangle $\Theta \subset \mathbb{R}^k$.*

To stress the dependency of $\mathcal{M}$ on the parameters $\theta \in \Theta$, we will often write $\mathcal{M}_\theta$. A PCRN $\mathcal{M}_\theta$ defines a Continuous Time Markov Chain [6,21] on D, with infinitesimal generator Q, where $Q_{\boldsymbol{x},\boldsymbol{y}} = \sum_{r_j \in \mathcal{R}} \{\alpha_j(\boldsymbol{x}, \theta) \mid \boldsymbol{y} = \boldsymbol{x} + \boldsymbol{v_j}\}$, $\boldsymbol{x} \neq \boldsymbol{y}$.

In the following, we will denote by $Path^\mathcal{M}$ the set of paths of a PCRN $\mathcal{M}_\theta$, which induces on it a probability measure $P_\theta = P_{\mathcal{M}_\theta}$, by the standard cylindric construction [1]. An element of $Path^\mathcal{M}$ is a cadlag function [5] $\boldsymbol{x} : [0, \infty) \to D$.

2.2 Signal Temporal Logic

Signal Temporal Logic (STL, [20]) is a discrete linear time temporal logic used to reason about the future evolution of a path in continuous time. Generally this formalism is used to qualitatively describe the behaviors of trajectories of differential equations or stochastic models. The temporal operators we consider are all time-bounded and it implies that time-bounded trajectories are sufficient to assess to verify every formula. The atomic predicates of STL are inequalities on a set of real-valued variables, i.e. of the form $\mu(\boldsymbol{X}):=[g(\boldsymbol{X}) \geq 0]$, where $g : \mathbb{R}^n \to \mathbb{R}$ is a continuous function and consequently $\mu : \mathbb{R}^n \to \{\top, \bot\}$. For our purposes, atomic predicates will be interpreted over a given PCRN $\mathcal{M}_\theta$.

Definition 2. *A formula $\phi \in \mathcal{F}$ of STL is defined by the following syntax:*

$$\phi := \bot \mid \top \mid \mu \mid \neg\phi \mid \phi \vee \phi \mid \phi\mathbf{U}_{[T_1,T_2]}\phi, \tag{1}$$

where μ are atomic predicates as defined above, and $T_1 < T_2 < +\infty$.

Eventually and globally modal operators can be defined as customary as $\mathbf{F}_{[T_1,T_2]}\phi \equiv \top\mathbf{U}_{[T_1,T_2]}\phi$ and $\mathbf{G}_{[T_1,T_2]}\phi \equiv \neg\mathbf{F}_{[T_1,T_2]}\neg\phi$. STL formulae are interpreted over the paths $Path^{\mathcal{M}}$ of a PCRN $\mathcal{M}_\theta$. We will consider two semantics: a boolean semantic [20], which given a trajectory $\boldsymbol{x}(t)$, returns either true or false, and a quantitative semantic [14], returning a real value capturing a notion of robustness of satisfaction.

Definition 3 (Boolean Semantics). *Given an element $\boldsymbol{x} \in Path^{\mathcal{M}}$, the boolean semantics $\models\subset (Path^{\mathcal{M}} \times [0,\infty)) \times \mathcal{F}$ is defined recursively by:*

– $(\boldsymbol{x}, t) \models \mu \iff \mu(\boldsymbol{x}(t)) = \top$

– $(\boldsymbol{x}, t) \models \neg\phi \iff (\boldsymbol{x}, t) \not\models \phi$

– $(\boldsymbol{x}, t) \models \phi_1 \vee \phi_2 \iff (\boldsymbol{x}, t) \models \phi_1 \text{ or } (\boldsymbol{x}, t) \models \phi_2$

– $(\boldsymbol{x}, t) \models \phi_1\mathbf{U}_{[T_1,T_2]}\phi_2 \iff$ *exists $t' \in [t + T_1, t + T_2]$ such that $(\boldsymbol{x}, t') \models \phi_2$ and for all $t'' \in [t, t')$ $(\boldsymbol{x}, t'') \models \phi_1$*

The boolean semantics describes if a trajectory satisfies or not an STL formula, but gives no information on how robust is this satisfaction. The quantitative semantics [4,14] fills this gap. The following quantitive semantics, given a STL formula ϕ and a trajectory $\boldsymbol{x}$, returns a real number $\rho(\phi, \boldsymbol{x}, t = 0)$ whose sign captures the truth value of the formula (positive if and only if true), and whose absolute value gives a measure on how robust is the satisfaction.

Definition 4 (Quantitative Semantics). *The quantitative satisfaction function $\rho : \mathcal{F} \times Path^{\mathcal{M}} \times [0,\infty) \to \mathbb{R}$ is defined by:*

– $\rho(\top, \boldsymbol{x}, t) = +\infty$
– $\rho(\mu, \boldsymbol{x}, t) = g(\boldsymbol{x}(t))$ *where g is such that $\mu(\boldsymbol{X}) \equiv [g(\boldsymbol{X}) \geq 0]$*
– $\rho(\neg\phi, \boldsymbol{x}, t) = -\rho(\phi, \boldsymbol{x}, t)$
– $\rho(\phi_1 \vee \phi_2, \boldsymbol{x}, t) = \max(\rho(\phi_1, \boldsymbol{x}, t), \rho(\phi_2, \boldsymbol{x}, t))$
– $\rho(\phi_1\mathbf{U}_{[T_1,T_2]}\phi_2, \boldsymbol{x}, t) = \sup\limits_{t' \in [t+T_1,t+T_2]} (\min(\rho(\phi_2, \boldsymbol{x}, t), \inf\limits_{t'' \in [t,t')} \rho(\phi_1, \boldsymbol{x}, t)))$

Boolean and quantitative semantics can be lifted from single trajectories to PCRN models, by leveraging the probability measure $P_\theta = P_{\mathcal{M}_\theta}$. More specifically, given a formula ϕ, we can extend the boolean semantics to define the satisfaction probability of ϕ as

$$P(\phi|\theta) \equiv P(\phi \mid \mathcal{M}_\theta) := P_\theta(\{\boldsymbol{x} \in Path^{\mathcal{M}} \mid (\boldsymbol{x}, 0) \models \phi).$$

Similarly, the quantitative semantics will map each trajectory of $Path^{\mathcal{M}}$ into a real number, thus defining a random variable R_ϕ over the reals:

$$P(R_\phi \in [a,b]) := P_\theta^{-1}(\boldsymbol{x} \in Path^{\mathcal{M}} \mid \rho(\phi, \boldsymbol{x}, 0) \in [a,b]),$$

of which we can compute the expected value or higher order moments (see [4] for more details):

$$\rho(\phi|\theta) := \mathbb{E}_{\mathcal{M}_\theta}[R_\phi].$$

2.3 Multi-objective Optimization Problem

Let us consider a vectorial function $f : \mathbb{R}^m \to \mathbb{R}^n$, $f(\boldsymbol{x}) = (f_1(\boldsymbol{x}), f_2(\boldsymbol{x}), \dots, f_n(\boldsymbol{x}))$ $\boldsymbol{x} \in \mathbb{R}^m$. We define an optimization problem associated to f as follows:

Definition 5. *An optimization problem associated to a vectorial function* $f(\boldsymbol{x}) = (f_1(\boldsymbol{x}), f_2(\boldsymbol{x}), \dots, f_n(\boldsymbol{x}))$ *consists in identifying the subset of the Domain of f such that:*
$$\boldsymbol{x}^* = argmin_{\{\sim_1, \dots, \sim_n\}} (f_1(\boldsymbol{x}), \dots, f_n(\boldsymbol{x}))$$
where $\sim_i \in \{\leq, \geq\}$ and the arg min *is referred to the partial order relation on $\mathbb{R}^n$ induced by $\{\sim_1, \dots, \sim_n\}$.*

Definition 6. *Given $\boldsymbol{x}, \boldsymbol{x}' \in \mathbb{R}^m$ we say that $\boldsymbol{x}'$ dominates $\boldsymbol{x}$ ($\boldsymbol{x} \preceq \boldsymbol{x}'$) respect to an optimization problem (Eq. 5) if $f_i(\boldsymbol{x}) \sim_i f_i(\boldsymbol{x}')$ for each $i = 1, \dots, n$ where $\sim_i \in \{\leq, \geq\}$ depending on if we are interesting in minimize or maximize f_i.*

Definition 7. *The Pareto optimal set $\mathcal{P}$ associated to a given optimization problem is the subset of the domain $\mathcal{D}$:*

$$\mathcal{P} = \{\boldsymbol{x} \in D | \nexists \, \boldsymbol{x}' \in D, \, \boldsymbol{x} \neq \boldsymbol{x}' \land \boldsymbol{x} \preceq \boldsymbol{x}'\} \tag{2}$$

Definition 8. *The Pareto frontier associated to a given optimization problem is the subset of the codomain of f described by:*

$$\mathcal{FP} = \{f(\boldsymbol{x}) \mid \boldsymbol{x} \in \mathcal{P}\},$$

where $\mathcal{P}$ is the pareto optimal set.

Basically speaking the Pareto optimal set is the best compromise we can obtain by maximizing or minimizing at the same time different functions on the same domain set. A point belongs to it if an improvement of an objective function causes another objective function to deteriorate.

3 Case Studies

We introduce now two case studies that will be used to illustrate our method and discuss the potential of multi-objective optimisation in this logic-based setting. We will consider two simple and well studied systems: a SIRS epidemic model and a Genetic Toggle Switch.

3.1   Epidemic Model: SIRS

We consider a SIRS epidemic model, introduced back in 1927 by W.O. Kermack and A.G. McKendrick [17], which is still widely used to simulate the spreading of a disease among a population. The population of N individuals is typically divided in three classes (though we will consider an extra one for vaccination):

- *susceptible* S, representing healthy individuals that are vulnerable to the infection.
- *infected* I, describing individuals that have been infected by the disease and are actively spreading it.
- *recovered* R and *vaccinated* V, modelling individuals that are immune to the disease, by having recovered from the disease (R), or by vaccination (V).

The basic SIRS model can be modified in many ways: considering vaccinations, introducing natural birth and death process, describing a latent phase of the disease, see [8] for an overview. We consider a version of the model described by the following set of reactions:

$$r_1 : S + I \xrightarrow{\alpha_1} 2I \qquad \alpha_1 = k_{si} \cdot \frac{X_s \cdot X_i}{N}$$

$$r_2 : I \xrightarrow{\alpha_2} R \qquad \alpha_2 = k_{ir} \cdot X_i$$

$$r_3 : S \xrightarrow{\alpha_3} R + V \qquad \alpha_3 = k_v \cdot X_s$$

$$r_4 : I \xrightarrow{\alpha_4} S \qquad \alpha_4 = k_{is} \cdot X_i$$

$$r_5 : R \xrightarrow{\alpha_5} S \qquad \alpha_5 = k_{rs} \cdot X_r$$

Here, r_1 describes the possibility that a susceptible gets the disease and becomes infected. The reaction r_2 models the recovery of an infected agent. The reaction r_3 describes the possibility that susceptible becomes directly a recovered by vaccination (k_v is the vaccination rate). Note that here the population V counts the total number of vaccinated people. Finally, r_4 describes the possibility that an infected individual recovers without gaining immunity, while r_5 models the loss of immunity of a recovered individual. Depending on the parameters $\theta = (k_{si}, k_{ir}, k_v, k_{is}, k_{rs})$, different behaviors of the disease could occur, such as the disease rapidly stops, or it becomes endemic, or there are periodic cycles of infections, and so on.

3.2   Genetic Toggle Switch

The Genetic Toggle Switch [17] is a regulatory genetic network composed by two genes that mutually repress each other. This is known to be a simple genetic circuity manifesting bistability and memory. It can be described by the following reactions:

$$r_1 : \emptyset \xrightarrow{\alpha_1} X_1 \qquad \alpha_1 = 1$$
$$r_2 : \emptyset \xrightarrow{\alpha_2} X_2 \qquad \alpha_2 = 1$$
$$r_3 : X_1 \xrightarrow{\alpha_3} \emptyset \qquad \alpha_3 = \frac{a_1 N^{b_1+1}}{N^{b_1} + X_2^{b_1}}$$
$$r_4 : X_2 \xrightarrow{\alpha_4} \emptyset \qquad \alpha_4 = \frac{a_2 N^{b_2+1}}{N^{b_2} + X_1^{b_2}}$$

where the reaction $\emptyset \to X_1$ means that the protein X_1 is created and $X_1 \to \emptyset$ means it is degraded. The model describes genes implicitly, by using Hill's kinetic rate functions, and lumps transcription and translation into a single step. Depending on the parameters, the systems can show two different behaviors: either a stable bistability or rapid switching between equilibria. In the first case the system will tend to stabilise for a long time in one of the two stable equilibria, (either $X_1 \geq X_2$ or $X_2 \geq X_1$). In the second case, the systems frequently switches from the state $X_1 \geq X_2$ to $X_2 \geq X_1$, and viceversa.

4   Problem Definition and Approach

In this paper we consider the following system design problem: given a PCRN $\mathcal{M}_\theta$, with h tunable parameters $\theta = (\theta_1, \ldots, \theta_h)$ and k STL formulae $\Phi = \{\phi_1, \ldots, \phi_k\}$, find a value θ^* of the parameters such that all formulae are satisfied "as much as possible" (i.e. their satisfaction probability is maximised). There is obviously a problem here, as we have k formulae that can represent conflicting objectives. Hence it may be impossible to maximise the satisfaction probability of all formulae at the same time. Moreover, different formulae can play a different role: we need to solve a *multi-objective* optimisation problem, that is we need to estimate the Pareto frontier. Additionally, we have two semantics for our logic, so we can decide to optimise either the satisfaction probability $P(\phi_j|\theta)$ or the robustness score $\rho(\phi_j|\theta)$. Note that both these quantities are *functions* of the model parameters θ. In the paper, we consider and compare three different approaches:

1. **Direct probability Approach (DpA):** solve directly the multi-objective problem associated to the maximization of the probability

$$P(\Phi|\theta) = (P(\phi_1|\theta), P(\phi_2|\theta), \ldots, P(\phi_k|\theta))$$

2. **Direct robustness Approach (DrA):** solve the multi-objective problem associated to the maximization of the robustness

$$\rho(\Phi|\theta) = (\rho(\phi_1|\theta), \rho(\phi_2|\theta), \ldots, \rho(\phi_k|\theta))$$

3. **Mixed Approach(MA):** combine the two approaches automatically switching among them.

Considering our goal, the DpA seems the natural choice. However, the quantitative semantics carries information about the satisfiability of a formula for each trace, plus additional information about robustness of satisfaction. As discussed in [4], when we average the robustness score over all trajectories, according to the distribution P_θ on trajectories, induced by a PCRN $\mathcal{M}_\theta$, we typically obtain a score which is positively correlated with the satisfaction probability. This means that, as a function of θ, typically satisfaction probability increases when robustness does (see for instance Fig. 2).

Furthermore, the robustness score typically carries more information in regions of the parameter spaces in which the satisfaction probability is flat, e.g. equal to zero or to one. In some cases, in fact, it happens that the probability vector is higher than zero only in a very tiny zone of the parameters space (this will be the case, for instance, in the Toggle Switch). The flatness of the objective vector function in a large area of the search space is a serious challenge for the optimization process, which will be forced to explore the objective space randomly and it is likely to remain stuck in such region. In these regions, however, the robustness score is typically non-constant, hence robustness is useful to guide the optimisation in these cases.

However, one can easily check that $\rho(\phi_i|\theta) > \rho(\phi_i|\theta') \not\Rightarrow P(\phi_i|\theta) > P(\phi_i|\theta')$. This implies that maximizing directly the robustness (DrA) could produce under optimal results. In fact, even if typically an increase in the average robustness corresponds to an increase in the satisfaction probability (Fig. 2), this may fail for some parameters θ and θ' such that $\rho(\phi_i|\theta) > \rho(\phi_i|\theta')$ and $P(\phi_i|\theta) < P(\phi_i|\theta')$.

This discussion leads us naturally to consider mixed strategies. In the MA approach considered in this paper, we have modified a genetic algorithm allowing the possibility to compare two designs based on the probability or on the robustness degree automatically. When two designs have the same probability with respect to the STL formulae, the algorithm will automatically switch to the robustness degree. Basically, this algorithm has the possibility to switch to robustness whenever it is stuck in a plateau of the satisfaction probability vector, which typically happens not only when all formulae are satisfied with probability zero, but also with probability one (which is part of the Pareto frontier). In this region, the robustness score leads us to pick the most robust θ^* satisfying the design problem.

Computing the Satisfaction Probability and the Robustness Scores.
The exact computation of the satisfaction probability or of the expected robustness, up to a fixed numerical precision ϵ is unfeasible, hence we will exploit statistical methods. More specifically, for each fixed θ, we sample N trajectories $\texttt{Traj} = \{x_1, x_2, \ldots, x_N\}$ of the PCRN using Gillespie's algorithm [18]. Then, for each STL formula ϕ_i of interest, we estimate $P(\phi_i|\theta)$ and $\rho(\phi_i|\theta)$ as

$$\hat{P}(\phi_i|\theta) = \frac{|\{x \in \texttt{Traj} | (x,0) \models \phi_i\}|}{N}; \qquad \hat{\rho}(\phi_i|\theta) = \frac{\sum_{j=1}^k \rho(\phi_i, x_j, 0)}{N}$$

Estimating statistically the functions to optimise is computationally feasible, but has the side effect of making their evaluation noisy, which must be taken into account in the optimisation phase. Here we try to reduce this effect by using a large number of simulations per point of the parameter space explored (here we used 500 runs per point, as a good trade off between noise and computational cost). An appealing alternative would be to rely on statistical regularization methods like Gaussian Processes-based emulation [22], which are typically an ingredient of active learning algorithm, like Pareto Active Learning (PAL, [25]).

The Multi-objective Optimisation Algorithm. The optimiser we have used is the NSGAII [13], a well known genetic algorithm largely used to solve multi-objective optimisation problems. It combines mutation and crossover operators to allow the creation of new points. A comparison procedure based on the Pareto dominance will give preference to points which dominate the largest number of other points, pushing effectively the optimization algorithm towards the Pareto frontier. We have slightly modified the comparison process: if two points have the same probability with respect to an STL formula, the comparison between them will be based on the robustness degree. Otherwise, the comparison is based on the probability. This strategy permits the optimization process to easily escape from plateaux of the objective function, as discussed previously.

5   Results

We discuss now two bi-objective optimization problems, one for each case study. The stochastic systems were simulated and the STL formulae verified with the U-check tool (see [7]).

5.1   Genetic Toggle Switch

In the Genetic Toggle Switch model described in Subsect. 3.2 there are 4 parameters, (a_1, a_2, b_1, b_2), which we assume that can take values in $[10^{-4}, 5]$, fixing the initial conditions to $(X_1(0), X_2(0)) = (1, 1)$. Depending on the parameters, the system can have two stable equilibria, typically one for which $X_1 > X_2$ and one for which $X_2 > X_1$. In particular, when the difference between X_1 and X_2 is above a certain threshold, then the system tends to stabilize in one equilibria, and the switching probability (i.e. spontaneously jumping to the other equilibria) will be low. Thus, we consider the two STL formulae:

$$\phi_1 := F_{[0,1000]}|X_1 - X_2| > 300 \tag{3}$$

$$\phi_2 := F_{[0,300]}G_{[0,50]}(X_1 > X_2) \wedge F_{[300,550]}G_{[0,50]}(X_1 < X_2). \tag{4}$$

The first formula describes a situation in which equilibria are clearly separated. This has as the side effect the tendency to stabilise the system into one equilibrium. The second formula, instead, describes a switch between equilibria, and hence it is in conflict with the previous goals. In particular, the higher is the difference between X_1 and X_2, the lower is the probability of ϕ_2. However, one may be interested in maximising the probability of both goals, to obtain a system with well separated configurations and the ability to explore them by the effect of noise. This is a typical use of randomness in cellular decision making, see for instance [2].

The bi-objective optimization problem we have considered is therefore:

$$\max_{a_1,a_2,b_1,b_2 \in [10^{-4},5]} \{P(\phi_1|\{a_1,a_2,b_1,b_2\}), P(\phi_2|\{a_1,a_2,b_1,b_2\})\} \qquad (5)$$

The three approaches described in Sect. 4 are compared in Fig. 1. It can be seen that the mixed (MA) approach performs best, while using only satisfaction probability (DpA) produces very bad results. This is caused by the fact that both STL formulae have a non-zero probability only in a small fraction of the parameter space, and the DpA approach is not able to reach this region. The direct robustness approach (DrA) also produce bad results. In this case the optimization concentrates itself on optimizing the robustness in the zone of the robustness degree space where the probability (at least for ϕ_2) is zero (See Fig. 1 (b)).

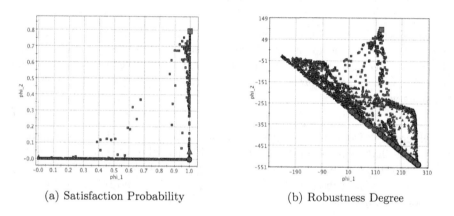

(a) Satisfaction Probability   (b) Robustness Degree

Fig. 1. Comparison of the three optimization approaches for the toggle switch. The red squared dots represent the MA approach, the blue triangular dots represent the DrA approach and the green circular dots are referred to the DpA. Chart (a) shows all designs, with larger dots representing the Pareto frontiers of the referred approaches. In (b) we have represented the same dots respect to the robustness degree space. (Color figure online)

5.2 SIRS Model

We consider now the SIRS model[1] of Sect. 3.1 and look for the best vaccination rate k_v to obtain the following behaviour for susceptible individuals

$$\phi_1 := G_{[20,40]} (S < 30 \wedge S > 10) \wedge F_{[40,60]} S > 50, \qquad (6)$$

---

[1] Rate coefficients are as follows: $k_{si} = 0.5$, $k_{ir} = 0.05$, $k_{is} = 0.1$, $k_{rs} = 0.05$, $k_v \in [0.08, 10]$. The vaccination rate is the only free parameter that is optimised.

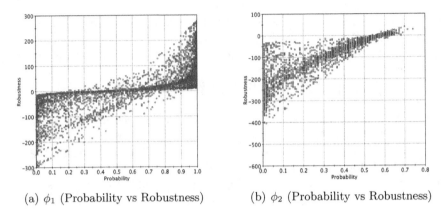

(a) ϕ_1 (Probability vs Robustness)   (b) ϕ_2 (Probability vs Robustness)

Fig. 2. Comparison of robustness and probability for the toggle switch for (a) ϕ_1 and (b) ϕ_2. We can see how the increase of the probability implies the increase of the robustness, even if this implication is not pointwise.

forcing susceptibles to be between 10 and 30 during time 20 to 40 and to get above 50 between time 40 and 60. We additionally want to constraint the number of vaccinations to be below 180 (say the amount of vaccines in storage, so that an higher consumption forces to buy new vaccines), which we enforce with the following goal

$$\phi_2 := G_{[0,250]} \, V < 180 \tag{7}$$

Differently from the Toggle Switch scenario, the three optimisation approaches have similar performances, as shown in Fig. 3. The reason is that the satisfaction probability of both ϕ_1 and ϕ_2 is non-zero in a large portion of the state space, and the correlation between probability and robustness is larger than for the toggle switch.

6   A Deeper Look at Robustness

In this section we focus our attention on the robustness score associated with a formula, starting from a potential criticism to multi-objective optimisation. Logic formulae can be combined, for instance by boolean connectives. Hence, if we have two formulae, ϕ_1 and ϕ_2, representing two potentially conflicting objectives, we can take their conjunction and solve the single objective optimisation for $\phi_1 \wedge \phi_2$, e.g. with the method of [4] which uses the robustness score. Even if the two approaches are not directly comparable considering that they solve different tasks (see. [12]), the second approach could be appealing for computational reasons.

However, consider the following scenario: let ϕ_1, ϕ_2 and $\psi := \phi_1 \wedge \phi_2$ such that $\forall \theta \in \Theta$, $\rho(\phi_1|\theta) \in [a,b]$ and $\rho(\phi_2|\theta) \in [c,d]$ where $a,b,c,d \in \mathbb{R}$ and $b \ll c$. Clearly $\forall \theta \in \Theta, \rho(\psi|\theta) = \min(\rho(\phi_1|\theta), \rho(\phi_2|\theta)) = \rho(\phi_1|\theta)$, hence optimizing the robustness, we would completely ignore $\rho(\phi_2)$. As a less artificial example,

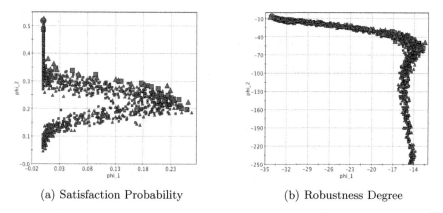

(a) Satisfaction Probability   (b) Robustness Degree

Fig. 3. Comparison of the three optimization approaches for the toggle switch. The red squared dots represent the MA approach, the blue triangular dots represent the DrA approach and the green circular dots are referred to the DpA. Chart (a) shows all designs, with larger dots representing the Pareto frontiers of the referred approaches. In (b) we have represented the same dots respect to the robustness degree space. Unlike the Toggle Switch example, the three approaches produce comparable results. (Color figure online)

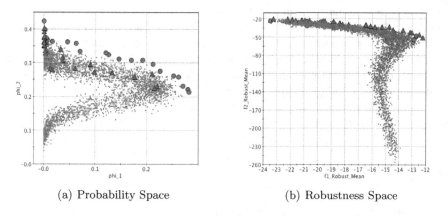

(a) Probability Space   (b) Robustness Space

Fig. 4. Comparison of the Pareto frontier or the SIRS model, obtained by maximizing the robustness and the probability are provided. The Pareto frontier obtained by maximizing the probability (green circular designs) dominates the Pareto frontier obtained by maximizing the robustness (blue triangular designs). (Color figure online)

we have compared these multi-objective and single-objective approaches applied to the maximization of the average robustness of the SIRS model (see Sects. 5.2 and 3.1) with respect to the formulae (6) and (7). The results (Fig. 5) show that the single objective problem tends to maximize the average robustness of the second formula, largely ignoring the first one.

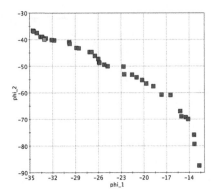

Fig. 5. Comparison of multi-objective optimisation of formulae (6) and (7) for the SIRS model. The blue boxes represent the pareto frontier relative to the maximization of the average robustness. The red box instead represents the result of the maximisation of the average robustness of the conjunction of the two formulae (reporting individual robustness scores). (Color figure online)

Indeed, these examples are paradigmatic of a more structural problem with the robustness score, namely its sensitivity to the scale of each atomic predicate. Essentially, a formula with more than one atomic predicate will combine their values in ways that may not be fully meaningful if the scale (domain of variability) of the two predicates is very different, as one may dominate the other. Sorting out this problem by a standardization of the predicates may be difficult, as their range of variability is unknown a priori. To circumvent this issue, at least in a system design perspective, we can de facto resort to multi-objective optimisation. Intuitively, at least for a suitable subclass of formulae, we can try to rewrite them so to obtain a boolean combination of temporal formulae with a single atomic predicate, and then define a suitable multi-objective problem on such subformulae. We plan to investigate the feasibility of this approach in a future work.

7   Conclusion and Future Works

Discussion. In this paper we investigated the problem of designing stochastic models of CRN with respect to behavioural goals expressed in temporal logic, with a multi-objective approach. We discussed several score functions, using satisfaction probability, robustness, or a combination of the two, and compared them for two different case studies: a toggle switch and a SIRS epidemic model. We also discussed the advantages of multi-objective approach to deal with robustness more consistently.

Future Work. The present work used a genetic algorithm to solve the multi-objective optimization problem. We have modified its comparison criterion in order to use a combination of satisfaction probability and robustness degree.

We plan to investigate the use of statistically more refined optimization methods to deal with noise in the estimation of the objective function due to the use of simulation, like Pareto Active Learning [25]. This will allow us to reduce the number of simulation runs per point of the parameter space, thus improving complexity.

References

1. Baier, C.: On algorithmic verification methods for probabilistic systems. Universität Mannheim (1998)
2. Balázsi, G., van Oudenaarden, A., Collins, J.J.: Cellular decision making, biological noise: from microbes to mammals. Cell **144**(6), 910–925 (2011)
3. Ballarini, P., Guerriero, M.L.: Query-based verification of qualitative trends and oscillations in biochemical systems. Theor. Comput. Sci. **411**(20), 2019–2036 (2010)
4. Bartocci, E., Bortolussi, L., Nenzi, L., Sanguinetti, G.: System design of stochastic models using robustness of temporal properties. Theor. Comput. Sci. **587**, 3–25 (2015)
5. Billingsley, P.: Convergence of Probability Measures. Wiley, NewYork (1999)
6. Bortolussi, L., Hillston, J., Latella, D., Massink, M.: Continuous approximation of collective system behaviour: a tutorial. Perform. Eval. **70**(5), 317–349 (2013)
7. Bortolussi, L., Milios, D., Sanguinetti, G.: U-check: model checking and parameter synthesis under uncertainty. In: Campos, J., Haverkort, B.R. (eds.) QEST 2015. LNCS, vol. 9259, pp. 89–104. Springer, Heidelberg (2015)
8. Brauer, F.: Compartmental models in epidemiology. In: Mathematical epidemiology, pp. 19–79. Springer (2008)
9. Chatterjee, K., Majumdar, R., Henzinger, T.A.: Markov decision processes with multiple objectives. In: Durand, B., Thomas, W. (eds.) STACS 2006. LNCS, vol. 3884, pp. 325–336. Springer, Heidelberg (2006). doi:10.1007/11672142_26
10. Coello, C.A.C., Lamont, G.B.: Applications of multi-objective evolutionary algorithms, vol. 1. World Scientific (2004)
11. Deb, K.: Multi-objective Optimization Using Evolutionary Algorithms, vol. 16. Wiley, New York (2001)
12. Deb, K.: Multi-objective optimization. In: Search Methodologies, pp. 403–449. Springer (2014)
13. Deb, K., Agrawal, S., Pratap, A., Meyarivan, T.: A fast elitist non-dominated sorting genetic algorithm for multi-objective optimization: NSGA-II. In: Schoenauer, M., Deb, K., Rudolph, G., Yao, X., Lutton, E., Merelo, J.J., Schwefel, H.-P. (eds.) PPSN 2000. LNCS, vol. 1917, pp. 849–858. Springer, Heidelberg (2000). doi:10. 1007/3-540-45356-3_83
14. Donzé, A., Maler, O.: Robust satisfaction of temporal logic over real-valued signals. In: Chatterjee, K., Henzinger, T.A. (eds.) FORMATS 2010. LNCS, vol. 6246, pp. 92–106. Springer, Heidelberg (2010)
15. Etessami, K., Kwiatkowska, M., Vardi, M.Y., Yannakakis, M.: Multi-objective model checking of Markov decision processes. In: Grumberg, O., Huth, M. (eds.) TACAS 2007. LNCS, vol. 4424, pp. 50–65. Springer, Heidelberg (2007)
16. Forejt, V., Kwiatkowska, M., Parker, D.: Pareto curves for probabilistic model checking. In: Chakraborty, S., Mukund, M. (eds.) ATVA 2012. LNCS, vol. 7561, pp. 317–332. Springer, Heidelberg (2012)

17. Gardner, T.S., Cantor, C.R., Collins, J.J.: Construction of a genetic toggle switch in escherichia coli. Nature **403**(6767), 339–342 (2000)
18. Gillespie, D.T.: Exact stochastic simulation of coupled chemical reactions. J. Phys. Chem. **81**(25), 2340–2361 (1977)
19. Henzinger, T.A., Jobstmann, B., Wolf, V.: Formalisms for specifying Markovian population models. In: Bournez, O., Potapov, I. (eds.) RP 2009. LNCS, vol. 5797, pp. 3–23. Springer, Heidelberg (2009)
20. Maler, O., Nickovic, D.: Monitoring temporal properties of continuous signals. In: Lakhnech, Y., Yovine, S. (eds.) FORMATS 2004 and FTRTFT 2004. LNCS, vol. 3253, pp. 152–166. Springer, Heidelberg (2004)
21. Norris, J.R.: Markov chains, Number 2008. Cambridge University Press (1998)
22. Rasmussen, C.E., Williams, C.K.I.: Gaussian Processes for Machine Learning. MIT Press, Cambridge (2006)
23. Swain, P.S., Elowitz, M.B., Siggia, E.D.: Intrinsic and extrinsic contributions to stochasticity in gene expression. PNAS **99**(20), 12795–12800 (2002)
24. Zhou, A., Bo-Yang, Q., Li, H., Zhao, S.-Z., Suganthan, P.N., Zhang, Q.: Multiobjective evolutionary algorithms: a survey of the state of the art. Swarm Evol. Comput. **1**(1), 32–49 (2011)
25. Zuluaga, M., Sergent, G., Krause, A., Püschel, M.: Active learning for multiobjective optimization. ICML **1**(28), 462–470 (2013)

Author Index

Printed in the United States
By Bookmasters